WRITING THE HISTORY OF THE MIND

Science, Technology and Culture, 1700–1945

Series Editors

David M. Knight
University of Durham

and

Trevor Levere
University of Toronto

Science, Technology and Culture, 1700–1945 focuses on the social, cultural, industrial and economic contexts of science and technology from the 'scientific revolution' up to the Second world War. It explores the agricultural and industrial revolutions of the eighteenth century, the coffee-house culture of the Enlightenment, the spread of museums, botanic gardens and expositions in the nineteenth century, to the Franco-Prussian war of 1870, seen as a victory for German science. It also addresses the dependence of society on science and technology in the twentieth century.

Science, Technology and Culture, 1700–1945 addresses issues of the interaction of science, technology and culture in the period from 1700 to 1945, at the same time as including new research within the field of the history of science.

Also in the series

Science and Spectacle in the European Enlightenment
Edited by Bernadette Bensaude-Vincent and Christine Blondel

Entropic Creation
Religious Contexts of Thermodynamics and Cosmology
Helge S. Kragh

British University Observatories 1772–1939
Roger Hutchins

William Crookes (1832–1919) and the Commercialization of Science
William H. Brock

Writing the History of the Mind

Philosophy and Science in France, 1900 to 1960s

CRISTINA CHIMISSO
The Open University, UK

Research for this book supported by the

Arts & Humanities
Research Council

ASHGATE

Published by
Ashgate Publishing Limited
Gower House
Croft Road
Aldershot
Hampshire GU11 3HR
England

Ashgate Publishing Company
Suite 420
101 Cherry Street
Burlington, VT 05401-4405
USA

Ashgate website: http://www.ashgate.com

British Library Cataloguing in Publication Data
Chimisso, Cristina
Writing the history of the mind : philosophy and science in France, 1900 to 1960s. – (Science, technology and culture, 1700–1945)
1. Philosophy of mind – History – 20th century 2. Philosophy – History – 20th century 3. Philosophy, French – 20th century 4. Philosophy and science – France – History – 20th century
I. Title
194

Library of Congress Cataloging-in-Publication Data
Chimisso, Cristina.
Writing the history of the mind : philosophy and science in France, 1900 to 1960s / Cristina Chimisso.
p. cm. — (Science, technology and culture, 1700–1945)
Includes bibliographical references.
ISBN 978-0-7546-5705-7 (alk. paper)
1. Philosophy of mind–History–20th century. 2. Philosophy–History–20th century. 3. Philosophy, French–20th century. 4. Philosophy and science–France–History–20th century. I. Title.

BD418.3.C46 2008
194–dc22

2007044797

ISBN 978 0 7546 5705 7

Printed and bound in Great Britain by MPG Books Ltd. Bodmin, Cornwall.

Contents

Series Editor's Introduction

There is a general perception that philosophy in general, and with it the history and philosophy of science and technology in particular, took a very different course in twentieth-century France compared to what happened in the English-speaking world. Alexandre Koyré, a giant in the Anglophone world, occupied, as Cristina Chimisso shows in this fascinating study, an equivocal and ambiguous place in the French establishment, neither an historian nor a philosopher. Writing with beautiful clarity, she examines through biographical and institutional history the particular developments that characterized French thinking in the first sixty years or so of the twentieth century. These years began with the Dreyfus affair and the polarization of society between the secular and the Catholic; and then came two World Wars, Nazi occupation and the loss of a progressive optimism in which evil was no more than the absence of good.

Indicating how philosophy, seen as intimately linked to science, was taught in France through its history, Cristina Chimisso illustrates the tensions between the history of philosophy and of science: the latter generally seen as progressive, the former timeless in the sense that 'classics' are never superseded. Some saw history as the laboratory of the philosopher; others argued about how far intellectual history depended upon historical and social situations. She contrasts the smooth careers and firm social position of those philosophers who moved from a Parisian *lycée* to the Ecole normale and on to prominent posts in academe, with the often Jewish or provincial background of many breaking into Parisian intellectual life through the new discipline of the history and philosophy of science, medicine and technology, and Helene Metzger's marginal place as a woman.

Her vivid biographical sketches, closely linked to reflections on the nature of the history of science, medicine and technology, bring to life a work demonstrating how personality, education, institutions and patronage have steered a particular national tradition in the history and philosophy of science. This stimulating study will be an example to any of us writing and thinking about intellectual history; and will make Anglophones look to French examples, not only for information but also for inspiration.

David Knight

Acknowledgements

I am very grateful to Arts and Humanities Research Council for awarding me a research leave grant, which enabled me to finish writing this book. The AHRC funds postgraduate training and research in the arts and humanities, from archaeology and English literature to design and dance. The quality and range of research supported not only provides social and cultural benefits but also contributes to the economic success of the UK. For further information on the AHRC, please see their website www.ahrc.ac.uk.

Many colleagues generously helped me at different stages of this project; I would like to thank in particular Nick Jardine, Martin Kusch, Warren Schmaus, Gad Freudenthal and Brian Alleyne. My gratitude also goes to the participants of the 2007 HOPOS conference in Paris and the 2006 BSHP conference in Cambridge (UK), both for their stimulating comments on my papers, which were to become part of the present book, for their own illuminating papers, and for the informal discussions we had about French history of philosophy and the history and philosophy of science.

Introduction

Historical Reconstructions and Disciplinary Boundaries

It is widely accepted that the French tradition of the philosophy of science significantly differs from its Anglo-American counterpart. This distinction has now been sanctioned by encyclopaedias of philosophy, which have separate entries for 'philosophy of science' and 'French philosophy of science', and by general works on the philosophy of science which also offer different accounts of these two traditions. In his entry on 'French philosophy of science' in the *Routledge Encyclopedia of Philosophy*, Gary Gutting, after an introduction in which he draws an ideal line linking Descartes, the Enlightenment, Auguste Comte, Pierre Duhem, Emile Meyerson and Henri Poincaré, proceeds to in-depth analyses of Gaston Bachelard, Georges Canguilhem, Michel Foucault and Michel Serres.[1] Similarly, Dominique Lecourt, in his French-language presentation of the 'French tradition' of philosophy of science, has emphasized its originality and its separation from logical positivism. Lecourt has outlined this tradition, which he rightly calls also 'the philosophical history of the sciences', by first mentioning Auguste Comte, Antoine Augustin Cournot and Condorcet, and then briefly presenting the twentieth-century protagonists, including Pierre Duhem, Henri Poincaré, Emile Meyerson, Abel Rey, Léon Brunschvicg and Alexandre Koyré. However, Lecourt's focus is firmly on 'the emblematic figure' of the 'French tradition of philosophy of science': Gaston Bachelard. Indeed, the most important representative of the philosophy of the life sciences, Georges Canguilhem, is presented as developing Bachelard's philosophy.[2] Bachelard and Canguilhem's 'historical epistemology' has indeed been the most renowned French philosophy of science, notably for its integration of history and philosophy.

Some critics, however, have challenged the received image of French philosophy of science as pivoting around historical epistemology and Bachelard's philosophy in particular. Some scholars have chosen to focus their attention on philosophers, such as Emile Meyerson, whose secondary place in the history of this discipline

---

1 Gary Gutting, 'French philosophy of science', in E. Craig (ed.), *Routledge Encyclopedia of Philosophy* (London: Routledge, 1998), available at http://www.rep.routledge.com/article/Q038 (accessed 9 October 2006). D. Vernant, in his *Encyclopédie philosophique universelle* entry on 'Epistémologie' (best translated as 'philosophy of science') distinguishes several traditions, including 'historical epistemology', which includes Koyré, Bachelard, Canguilhem and also Feyerabend, Kuhn and Lakatos; a tradition dedicated to the reflection on the human sciences (Foucault and Piaget); and another whose protagonists focused on logic and analysis of language (Russell, Wittgenstein, Carnap, Reichenbach, Quine, Popper): *Encyclopédie philosophique universelle, vol. 2: Les notions philosophiques* (ed. Sylvain Auroux; Paris: Presses Universitaires de France, 1990), p. 813.

2 Dominique Lecourt, *La philosophie des sciences* (Paris: Presses Universitaires de France, 2001).

seems partly to be a consequence of Bachelard's opposition to their philosophies.[3] Critics have also challenged the clear-cut distinction between French and Anglo-American philosophy of science. Recently, Anastasios Brenner has argued that the tradition of logical positivism, and indeed of current Anglo-American philosophy of science, does not only have Austrian roots, but also French ones, notably in the conventionalism of Pierre Duhem and Henri Poincaré. Brenner does not reject the opposition between the two styles of philosophy of science, but rather argues that both have been practised in France. Indeed, he reads French philosophy of science after the First World War, which includes Bachelard and Koyré, and corresponds to the 'classical' image of French philosophy of science, as an attempt to go beyond conventionalism.[4] Projects like these are very important because they present a more complex image of the approaches to science that French scholars developed in the last century. However, they do not deny the existence, and indeed the dominance in France, of a specific tradition in philosophy of science that intimately connects history and philosophy.

My own project here is to present a new account of the intellectual tradition that culminated with Gaston Bachelard's and Georges Canguilhem's versions of historical epistemology. I do believe that this tradition, although it is by no means the only one worthy of attention in the rich philosophical landscape of twentieth-century France, is nonetheless of crucial importance and great originality. My narrative does not contradict any of those mentioned, but differs from them in at least two respects. The first is chronological. In English-language criticism, Bachelard's and Canguilhem's philosophies have often been presented as beginnings, for the general interest has been in the philosophies which developed in the last part of the twentieth century. The reception of Bachelard in the 1970s, especially through the translations of Lecourt's works, linked his philosophy to those of Canguilhem, Foucault and Althusser.[5] Moreover, almost symbolically, Bachelard and Canguilhem are the topics of the first chapter of Gary Gutting's volume on Michel Foucault, published in 1989.[6] By contrast, in the present book, Bachelard and Canguilhem occupy the last chapter, as my story starts in the first decades of the twentieth century.

Secondly, the main characters of my story partly differ from those found in accounts of French philosophy of science, simply because I do not aim to narrate the history of that illustrious tradition. It is not only the case that I will not aim to discuss all French philosophers of science. Other scholars, who are not considered

---

3 This, for instance, is Bernadette Bensaude-Vincent's current project as I understand it. Frédéric Fruteau de Laclos and Eva Teklès-Klein are also part of this effort to offer a new and better evaluation of Meyerson's philosophy as well as a more accurate account of his life and work based on archival work.

4 Anastasios Brenner, *Les origines françaises de la philosophie des sciences* (Paris: Presses Universitaires de France, 2003), pp. 99–109.

5 See Dominique Lecourt, *Marxism and Epistemology: Bachelard, Canguilhem and Foucault* (trans. Ben Brewster; London: NLB, 1975). This volume includes the translations of two French works: Dominique Lecourt, *L'épistémologie historique de Gaston Bachelard* (Paris: Vrin, 1969); and Dominique Lecourt, *Pour une critique de l'épistémologie* (Paris: Maspero, 1972).

6 Gary Gutting, *Michel Foucault's Archaeology of Scientific Reason* (Cambridge: Cambridge University Press, 1989).

philosophers of science, will also play important roles. The identity and boundaries of a discipline are always controversial, and it is not my intention here to attempt to define either. Indeed, in my account, disciplinary boundaries will appear blurred, as indeed they were in the milieu in which the scholars whom I shall discuss worked. It is already revealing that the central characteristic of a large part of French philosophy of science is its engagement with history. Indeed, many scholars have been classified in different ways: is Koyré a historian or a philosopher? Is Canguilhem's doctrine historical epistemology or rather epistemological history? Hélène Metzger may be known as a historian of chemistry, but she also made important contributions to philosophy. On the other hand, Lucien Lévy-Bruhl is generally excluded from histories of philosophy of science, and this is not surprising, for he did not engage with science. He is nowadays remembered as the author of volumes on traditional societies as described by ethnological reports, as the founder of the Institut d'Ethnologie at the University of Paris, and more rarely as a historian of philosophy, despite his chair in History of Modern Philosophy and his publications in this branch of philosophy. However, his theory of primitive mentality[7] was widely discussed and employed by historians and philosophers of science, and his questions were intimately connected with theirs. Similarly, Léon Brunschvicg, who was also a professor of History of Modern Philosophy, generally occupies a rather small place in accounts of philosophy of science, despite the great importance that his philosophy had for his student Bachelard.

Writing the History of the Mind: A Set of Projects across Disciplines

Rather than a history of a discipline, this book is the history of a set of projects which were aimed at investigating the mind through the examination of texts. The texts were generally part of European intellectual history, although Lévy-Bruhl chose to employ ethnological reports as his primary sources. The underlying assumptions that united these projects were that the mind could not be studied *a priori*, and that ways of thinking were different in different civilizations. As a consequence, history was as a rule an essential component of research. Past philosophy and past science were expected to reveal worldviews and mental processes that differed from current ones. These projects should not however be seen as a unified programme; in fact, beyond these shared aims, individual research agendas differed. Indeed, philosophically oriented scholars, such as Brunschvicg, in the eyes of some historians only 'used' history in order to demonstrate their thesis. By contrast, the aim of scholars such as Metzger and Koyré was principally the accurate reconstruction of past theories, practices and ways of thinking. However, any clear-cut distinction between philosophers and historians would be misleading. Indeed, a scholar such as Abel Rey could contrast his historical approach to Brunschvicg's philosophical one, only for Metzger to class both as philosophical, in contrast with her own historical work.

---

7 For the sake of simplicity, I translate the French term '*mentalité*' as 'mentality'. What French scholars intended by '*mentalité*' will become clear through the examination of their work.

It is more accurate to think of these projects as a spectrum, in which the relative importance of history and philosophy varied.

Moreover, even though my focus will be on historians of philosophy and historians of science, they were certainly not alone in studying the mind through history, nor indeed in aiming at writing the history of the mind. Although nowadays the scholars sharing this project are seen as belonging to different disciplines, and almost different worlds, most of them were in close contact. As a consequence, scholars who would normally be excluded from the histories of these disciplines, will in fact appear as interlocutors of the main characters of this book. Among them, there will be the sociologists Paul Fauconnet and Marcel Mauss. Durkheim himself, although he died before the end of the First World War, will be seen comparing his own project to that of Lévy-Bruhl. Moreover, we shall observe psychologists such as Henri Piéron engage in discussions with philosophers and historians. We shall also encounter Lucien Febvre, the historian of mentalities, not only in his important role at Henri Berr's Centre de synthèse, of which many of the scholars discussed here were members, but also as he collaborated with Abel Rey on the publication of the first volume of the *Encyclopédie française.* As the general editor, Febvre gave Rey the responsibility of writing the first chapter on the *outillage mental*, or mental tool. Psychoanalysis will appear in the discussion of the philosophy of Gaston Bachelard, who made a very personal use of it in order to understand the ways of reasoning of past intellectuals and in order to help the mind to become more scientific.

The combination of history and philosophy in the study of the mind, and indeed the idea of a history of the mind, were not new to philosophy, nor to French philosophy in particular. The most prominent French model of intellectual development in history was that of Auguste Comte (1798–1857); and indeed the Enlightenment had already provided examples, notably Condorcet's *Esquisse d'un tableau historique des progrès de l'esprit humain* (1795).[8] Twentieth-century scholars, including two as different as Emile Bréhier and Georges Canguilhem, often used the metaphor of history as the laboratory, or the experimental method, of the philosopher. The latter cited several past supporters of the idea of the use of history for constructing a theory of the human mind and a theory of knowledge, including Marie Jean Pierre Fleurens (1794–1867), who expressed it in his obituary of Georges Cuvier, and Pierre Laffitte, in his inaugural lecture at the Collège de France (1892).[9] Other examples could be added, such as, outside France, Giambattista Vico, but, although obviously I shall occasionally refer to pre-twentieth-century philosophers, these are outside the scope of the present book. Moreover, I think that the novelties of the twentieth-century projects of history of the mind should not be obliterated by their inevitable links to previous traditions. Nor should we forget that, despite the apparent similarities of their projects with Comte's, most scholars studied in this book held strong anti-

---

8 Caritat, Marie Jean Antoine Nicolas, Marquis de Condorcet, *Esquisse d'un tableau historique des progrès de l'esprit humain. Texte revu et présenté par O.H. Prior. Nouvelle édition présentée par Yvon Belaval* (Paris: Vrin, 1970 [1795]).

9 Georges Canguilhem, *Etudes d'histoire et de philosophie des sciences concernant les vivants et la vie* (Paris: Vrin, 1994 [1968]), p. 12.

positivist convictions, were on the whole opposed to laws of progress, and regarded intellectual history as an open-ended process.

From the History of Philosophy to the History and Philosophy of Science

My narration of a set of projects, which, in the twentieth century, were aimed at studying the history of the mind by combining history and philosophy, will start by considering its roots in the history of philosophy. The study of the mind was traditionally the realm of philosophy, and the branch of philosophy most devoted to history was the history of philosophy. It should not then be surprising that these projects should be linked to the history of philosophy. However, an analysis of methodological pronouncements and discussions will reveal a tension between the most accepted and common practices in the history of philosophy in the twentieth century, and the works that were going to be crucial for the history of the mind: those of the professors of History of Modern Philosophy, Lucien Lévy-Bruhl and Léon Brunschvicg. The roots of their projects were indeed in the history of philosophy, but their ways of considering past texts set them apart from many of their fellow historians of philosophy. In their different ways, the study of the mind became the main objective for both of them. As a consequence, they read texts not primarily for their immediate contents, but for the way of reasoning that they revealed. Moreover, they both believed that there exist different ways of reasoning, depending on era and civilization.

Their choices and views had two consequences. The first was that the content of a text was not taken at its face value, but rather as inevitably relative to a cultural context. The other consequence was that philosophical texts lost the special status which most of Lévy-Bruhl and Brunschvicg's colleagues attributed to them. For the majority of historians of philosophy, including Emile Bréhier, Victor Delbos, Léon Robin, Henri Bergson, Etienne Gilson and Jean Wahl, past philosophical texts were still relevant to the modern reader, even if they had been written centuries earlier; in this respect, they were different from scientific texts, which were superseded by more recent ones. By contrast, if the main aim of the reader of a text is to find out the type of mentality that produced it, then, in principle, any texts could be used. Lévy-Bruhl's and Brunschvicg's differences with the historians of philosophy should not, however, be seen as a simple opposition. They also practised the history of philosophy in the more 'mainstream' manner, and it could not have been otherwise, given the academic positions that they came to occupy. On the other hand, even their colleagues who more openly defended the timeless nature of philosophical texts and advocated a direct reading of them, such as Léon Robin, nonetheless dedicated much of their work to historical and philological research. The inevitable complexity of positions and practices should not make us miss the importance of the difference between Lévy-Bruhl's and Brunschvicg's ways of evaluating the role of texts and their aims, on the one hand, and those of most of their colleagues, on the other. Lévy-Bruhl and Brunschvicg appeared to regard past texts not as a heritage to preserve and keep alive, but as documents of ways of reasoning which were no longer current.

Not surprisingly, Lévy-Bruhl's and Brunschvicg's legacies were especially felt outside the history of philosophy. Lévy-Bruhl's theory of primitive mentality had a tremendous impact on many disciplines, and became a common reference in the history and philosophy of science. Hélène Metzger referred to his theory of mentality in order to elaborate her own concept of the mental *a priori*, just as Abel Rey did in his presentation of the *outillage mental*, and Brunschvicg in his analysis of the dawn of mathematics. On the other hand, Bachelard's reception of Brunschvicg's philosophy was crucial in the development of his own doctrine, and in the development of historical epistemology.

Academic and Intellectual Spaces

The links between the scholars presented in this book will not only be connections between disembodied ideas. These scholars shared not only ideas, but also places and time. They were not isolated intellectuals, but rather busy academics who lectured, organized conferences, gave and attended talks, and supported younger scholars' careers; indeed it would be anachronistic to seek connections between them by only examining what they read. Real and virtual places of work and discussion are central in this book. My profile of the history of philosophy will start with the professors of History of Philosophy at the Sorbonne and the Collège de France. I shall also examine academic history of philosophy from the students' point of view: how much history of philosophy did they have to study, and how much was on offer? I shall also evaluate the strength of the history of philosophy in terms of university chairs, doctoral dissertations and intellectual prestige. The two major philosophy journals, the *Revue philosophique* and the *Revue de métaphysique et de morale* were extremely important not only for philosophers, and indeed social scientists and other intellectuals, but also for historians of philosophy in particular. Incidentally, Lévy-Bruhl edited the former, while Brunschvicg was part of the 'inner circle' behind the latter. In order to investigate the methodology of the history of philosophy, I shall adopt the point of view of a hypothetical reader of these journals, as well as that of a participant in chosen sessions of the international conference of philosophy, organized by the *Revue de métaphysique et de morale* group. This group was also behind the Société française de philosophie, which provided an invaluable arena where the most important intellectuals discussed a variety of topics and questions. I shall rely on the transcripts of chosen discussions in order to present the different positions regarding history, the study of the mind, and the value of past texts.

Whereas philosophy, including the history of philosophy, had a very clear academic space in the twentieth century, the history of science did not. The space for philosophical history of science at the Sorbonne was created by Abel Rey in the early 1930s, in particular with the foundation of the Institut d'histoire des sciences et techniques, and the creation of related courses, examinations and diplomas. An extra-academic organization in which the history of science found an important home was the Centre de synthèse. Abel Rey, as a close associate of the general director Henri Berr, played an important role in it, which included the direction of one of its units, that of 'general synthesis', and co-editing the Centre's journal, the *Revue de synthèse*. Another unit, the Section d'histoire des sciences, directed by Aldo Mieli, was dedicated to the history of science. It was there that its administrator Hélène Metzger delivered her methodological and philosophical papers, and discussed her conception of the mind and its history. The minutes of the talks that took place in the Section d'histoire des science were published in Mieli's journal *Archeion*; they give us a vivid image of its members' different, and often conflicting, points of view. It was Metzger who introduced to the Section an historian of philosophy and religious ideas who was going to become an extremely important historian of science: Alexandre Koyré.

Scholars' Social Positions and their Philosophical and Political Ideas

Since discussions and interactions between scholars are central to my approach, I shall describe the institutions and groups in which these interactions took place. I shall also, however, consider the individual scholars' positions, and analyse the places that they occupied in society and in academia. Is it significant that most of them seem not to correspond to the average French academic? I shall discuss the relative marginality of several of them, although their social or academic 'marginality' will have a different meaning depending on individual cases. At times, it might appear paradoxical, for instance when applied to such important academics as Brunschvicg. And yet, his social background, along with that of Lévy-Bruhl, was rather less privileged than those of his colleagues. In the case of Metzger, her gender played an important role in keeping her at the margins of academia, while their foreign nationalities had different effects on the lives and careers of Mieli and Koyré. Was it significant that Lévy-Bruhl, Brunschvicg, Metzger and Koyré all had Jewish backgrounds? The organizers of the spaces analysed in this books were also Jewish, including Henri Berr, Aldo Mieli and the founder of the *Revue de métaphysique et de morale*, Xavier Léon. These scholars lived through the aftermath of the Dreyfus affair, and the older among them through the affair itself. Indeed, some of them appear to be linked by it; for instance, Lévy-Bruhl successfully asked Koyré to translate the Schwarztkoppen papers, which proved the innocence of Alfred Dreyfus.

The political and ethical beliefs of the main scholars discussed will also receive due attention, especially in their potential connections with their philosophies. This book, however, covers a relatively long period, and some scholars (e.g., Lévy-Bruhl, Brunschvicg, Metzger, Rey) never saw the end of the Second World War, whereas Koyré and Bachelard died in 1964 and 1962, respectively, and many of their major works were published after the war. Canguilhem's life stretched from 1904 to 1995, and his most important academic activities and publications took place from the Second World War onwards. The 'historians of the mind' active during the Third Republic appeared to embody the ideal type of the academic of that period: secular and republican. However, at the beginning of the book, I shall look a little more closely at the ideas of the historians of philosophy, who after all did not seem unanimously to share the secular and republican convictions of their colleagues: Lévy-Bruhl, a socialist, and Brunschvicg, who was close to the Radical party. It is difficult not to see a connection between Lévy-Bruhl, Brunschvicg and Rey's secularism[10] and progressive ideas, on the one hand, and their view of modernity as the era of science and rationality as opposed to a mystical and superstitious past, on the other. It is not by chance that Catholic scholars, including Lévy-Bruhl's own student, Etienne Gilson, defended the enduring relevance of a way of approaching reality that the secular scholars represented as primitive or at least contrary to the rational and scientific spirit. However, Metzger, who described herself as 'almost socialist', appeared to ascribe some role to 'primitive' ideas and indeed to religion in the development of scientific theories. The younger generation of scholars who were active beyond the Third Republic not surprisingly had a different relationship with the political landscape. The cosmopolitan Koyré, who linked metaphysical, religious and scientific ideas more than Metzger did, did not seem to share the political commitment of his French colleagues. Bachelard appeared philosophically linked to the projects elaborated during the Third Republic, but he took up his post at the Sorbonne at the very end of it, in 1940. After the Second World War, France and the world greatly changed, and so did the role of intellectuals. Bachelard, and especially his successor Georges Canguilhem, did not appear to be 'new intellectuals', but obviously nor could they play a social role similar to that of their Third Republic predecessors, nor perhaps could their philosophies have had the same links with contemporary political events.

Plan of the Work

I shall use all these different perspectives to present a unified narrative: I shall present the development of a group of projects of the history of the mind, and at the same time examine the individual scholars' ideas; I shall consider institutions and groups, and at the same time evaluate these scholars' social and academic positions. In Chapter 1, I shall offer a profile of the history of philosophy and its practitioners in France in the first decades on the twentieth century. The professors of the history of philosophy in these institutions had great influence on the future not only of this speciality, but of philosophy in general, and an analysis of their education, social and religious background will offer an interesting snapshot, especially in order to evaluate Lévy-Bruhl and Brunschvicg. Chapter 2 will present an overview of the historians of philosophy's conceptions of their discipline, by examining their contributions to journals, conferences and the Société française de philosophie. In Chapter 3, I shall focus on the work of Lévy-Bruhl and Brunschvicg, showing that, on the one hand, their roots in the history of philosophy were very important, but

10 For the sake of simplicity, I shall use the term 'secularism' throughout the book, including when it translates the term '*laïcité*', which does not have a perfectly correspondent English term, but that generally refers to the secular character of the state, and to the principle of separation of state and church. My analysis of the doctrines of the philosophers under study will hopefully provide a clearer explanation of their different conceptions of *laïcité* than any single term would. The period under study in the present book was crucial in the history of secularism in France from a political point of view, as the law that separated state and church was passed in 1905. The literature on *laïcité* is immense. Maurice Barbier, *La laïcité* (Paris: L'Harmattan, 1995), has a good bibliography, arranged by theme.

that, on the other, they had methodological disagreements with their colleagues. Chapter 4 will present the places in which the type of history of science connected with the project of writing the history of the mind flourished, sometimes against the odds. I shall dedicate particular attention to the Centre de synthèse, but also to the Sorbonne. The epistemological and historiographical views of the Centre's director, Henri Berr, and of his associates Mieli and Rey, will show the tensions between different conceptions of the aims and methods of the history of science. Chapter 5 will be focused on two members of the Centre de synthèse who were among the main protagonists of the set of projects that I analyse in this book: Hélène Metzger and Alexandre Koyré. Chapter 6 will be dedicated to Gaston Bachelard, who originally continued the project of the study of the mind through the study of the history of science, and to Georges Canguilhem, whose further elaboration of that project changed some of its fundamental aspects.

Chapter 1

The History of Philosophy in the First Decades of the Twentieth Century: The Spaces and the Students

The Identity of the History of Philosophy

The task of writing a history of the mind seems perfectly suited for historians of philosophy: the study of the mind has traditionally been within the realm of philosophy; among philosophers the most historically inclined are the historians of philosophy. However, at the beginning of the twentieth-first century it would be difficult to find historians of philosophy who would thus describe their work. Indeed, in English-speaking countries, historians of philosophy are somewhat hard to find at all, and, outside specific events, such as the annual meeting of the British Society for the History of Philosophy, it is unlikely to find them in great numbers. A department of philosophy without a historian of philosophy would not raise an eyebrow; likewise, nobody would be surprised if scholars working on Machiavelli are considered historians, or if those working on Foucault are in French departments.

France in the first half of the twentieth century offered a strikingly different picture. Philosophy as a whole played a crucial role both in secondary and higher education, and the history of philosophy was considered a very important part of philosophy in higher education and research. The importance of the history of philosophy, and of the role that some historians of philosophy played in the development of the intellectual tradition that combined the study of history, philosophy and science, have not received due attention. This tradition, which has become almost synonymous with French-style history and philosophy of science, has rarely been connected with the practice of studying philosophy historically and with the history of philosophy as exegesis of texts. However, many projects in this tradition, and in particular those aimed at writing the history of the mind, did have their roots in the work of historians of philosophy, in particular of Lucien Lévy-Bruhl and Léon Brunschvicg. These two scholars, however, brought into the history of philosophy approaches that many of their colleagues resisted. In order to examine some important research programmes, which resulted in original ways of approaching the history of science and in new epistemologies, it is necessary both to understand the profile of the history of philosophy in early twentieth-century France, and the ways in which some unorthodox (from the history of philosophy's point of view) approaches managed to influence future research.

Several philosophers whom I shall discuss in this book nowadays are rarely presented as historians of philosophy (although they are by French critics). Indeed,

I have sometimes encountered bemusement at my use of this professional and disciplinary category. Needless to say, I believe that it is historically and intellectually accurate to have such category. Having said this, it is by no means my intention to suggest that the history of philosophy had clear-cut boundaries. In fact, one of the main aims of the present book is to show that disciplinary boundaries were fluid, and that scholars seen nowadays as belonging to different disciplines in fact shared institutional and intellectual spaces. Their diverse approaches and methods often did not translate into a separation in terms of teaching, conferences and journals. First of all, what was called philosophy in France in the first decades of the twentieth century was much broader than for instance the field implied by the benchmarks of the United Kingdom's Quality Assurance Agency.[1] Metaphysical and epistemological research took place alongside philological studies, the development of sociological and ethnological approaches, as well as the elaboration of experimental psychology and the study of the history of science. Notwithstanding the creation of new disciplines, namely sociology, ethnology, experimental psychology and general history of the sciences, their practitioners had still not only common philosophical roots, but also strong intellectual, institutional and even personal links. Moreover, higher education maintained these disciplines in close contact in terms of curriculum.

The practitioners of the emergent disciplines also shared important intellectual spaces with the philosophers, including journals and learned societies. In the early twenty-first century, we are used to a very high number of learned societies and journals; even in universities in English-speaking countries, where the history of philosophy is at best struggling, there are societies and journals solely dedicated to this subject. The expansion of tertiary education, already a reality in the period under study,[2] has gathered unparalleled momentum in the last few decades. It would be anachronistic to expect the same number of journals and societies dedicated to specific branches of philosophy or other disciplines in the first decades of the twentieth century as there are nowadays. However, what appears to be a lesser degree of specialization of journals, societies and indeed teaching and research, was not only the result of a smaller number of practitioners and students; it was also one of the most important characteristics of this intellectual field. The history of these disciplines, which branched out of philosophy, partly explains their closeness to philosophy and to each other in the period under study, when their respective identities were in many cases not completely formed. On the other hand, the practitioners of the new disciplines needed to mark their differences from the philosophers, in order to define their own methods. Moreover, the fluidity of disciplinary boundaries was rooted in

---

1 See the Quality Assurance Agency for Higher Education website: http://www.qaa.ac.uk/academicinfrastructure/benchmark/honours/philosophy.asp (accessed 18 April 2006).

2 University officials already regarded the expansion of higher education as almost out of control in the 1930s: in his report on the academic year 1933–34, the Dean of the Faculty of Letters of the University of Paris warned students about their choice of studying humanities in increasingly great numbers (Faculté des Lettres, 'Rapport annuel du Doyen: année scolaire 1933–34', *Annales de l'Université de Paris* 10/5 [1935]: 385–404 [385]). That year, 9,965 students had enrolled in the Faculty of Letters, compared with 4,422 in 1924–25 (Faculté des Lettres, 'Rapport annuel du Doyen: année scolaire 1925–26', *Annales de l'Université de Paris* 2 /4 [1927]: 295–309 [296]).

the training of French students. French pupils were exposed to what were considered key disciplines throughout their schooling. This was no accident: the ideal of *culture générale* was fundamental to the educational ideals of the time, for the knowledge of a variety of disciplines was seen as the only way to form harmonious minds and good French citizens.[3]

Despite its blurred boundaries, and its fluid development, not only philosophy, but also the history of philosophy, can be described and evaluated in the first half of the twentieth century. There are indicators that can give us a clear idea of its profile: chairs and doctoral theses at the University of Paris, publications in academic journals, discussions at the Société française de philosophie, and views of contemporaries. In order to understand the history of philosophy, and philosophy as a whole, it is crucial to grasp their institutional profiles at this stage in their development. By contrast with the nineteenth century, in the twentieth century philosophy was fully institutionalized, and enjoyed a high profile in higher education. This situation was relatively new for philosophers; as Jean-Louis Fabiani reminds us, the great nineteenth-century philosophers, such as Auguste Comte, Antoine Cournot and Charles Renouvier, never taught philosophy; and indeed they had a scientific rather than philosophical training.[4] In the period under study here, philosophers were overwhelmingly philosophy teachers, either in secondary or in higher education, and often both during their lifetime. Indeed, many of them had a strong interest in education, which they expressed in publications and by participating in the running of the education system.[5]

I shall begin by briefly outlining the role of history of philosophy in *lycées*, and at the University of Paris, by looking at chairs and doctoral dissertations. The examination of the social and educational background of the Parisian professors

---

3 The literature on *culture générale* is vast. Here I will only mention two books by scholars whom the reader will encounter again in the present book: Célestin Bouglé, *Humanisme, sociologie, philosophie. Remarques sur la conception française de la culture générale* (Paris: Hermann, 1938); and Léon Brunschvicg, *Un Ministère de l'Education Nationale* (Paris: Plon, 1922). Both authors presented the ideal of comprehensive and integrated knowledge as necessary to the intellectual and moral development of students. Pierre Bourdieu has denounced the ideal of *culture générale* as a means of social reproduction, and presented it in less flattering terms as 'the art of being able to mobilize instantly all available resources and to get the most out of them': Pierre Bourdieu, *The State Nobility* (trans. Lauretta C. Clough; Cambridge: Polity Press 1998 [1989]), p.88. I have discussed *culture générale* elsewhere: see Cristina Chimisso, *Gaston Bachelard: Critic of Science and the Imagination* (London: Routledge, 2001), ch. 2.

4 Jean-Louis Fabiani, *Les philosophes de la République* (Paris: Editions de Minuit, 1988), p. 28.

5 A few examples among many: Gustave Belot, after teaching in *lycées* (replacing Lucien Lévy-Bruhl at the *lycée* Louis-Le-Grand) was rector of the Académie de Paris and director of secondary teaching (Archives Nationales, AJ/16/952); Dominique Parodi was Inspecteur général de l'Instruction publique (Anonymous, 'Chronique', *Annales de l'Université de Paris* 5/3 [1930]: 417–21 [419]; André Lalande, 'Dominique Parodi', *Revue de métaphysique et de morale* 60 [1955]: 341–51 [342]); Victor Delbos was member and then president of the Société d'études pour les questions d'enseignement secondaire (Christophe Charle, *Dictionnaire biographique des universitaires aux XIXe et XXe siècles, vol.1: La Faculté des Lettres de Paris (1809–1908)* [Paris: CNRS, 1985], pp. 58–9).

of the history of philosophy will provide food for some very preliminary thoughts about the possible link between one's background and one's view of history and of the mind.

Studying the History of Philosophy in Paris

The Uncertain Role of the History of Philosophy in Lycée *Syllabi*

As Léon Brunschvicg put it, university philosophy students were 'no novices' in this discipline; and indeed he and Henri Bergson, whom he quoted in his support, attached great importance to their students' early philosophical education.[6] As a rule their students would have had intense philosophy training in the last year of their secondary education, that is in the year of preparation for their *baccalauréat*.[7] The role of the history of philosophy in secondary education syllabi evolved from the time when Bergson, Brunschvicg and their contemporaries were *lycée* pupils, to the time when they were *lycée* teachers themselves, and university professors later on. In the 1870s the philosophy syllabus covered the following areas: psychology, logic, ethics, the history of philosophy and philosophical texts. The importance accorded to psychology and the history of philosophy was a legacy of Victor Cousin (1792–1867). Knowledge of the history of philosophy was crucial to Cousin's philosophical eclecticism, which was aimed at combining the 'essential and true' elements of all philosophical systems, and as a consequence required a solid knowledge of these systems.[8] In 1902, metaphysics and aesthetics were introduced, whereas the history

6 Léon Brunschvicg, 'La philosophie dans l'enseignement supérieure à Paris', *Annales de l'Université de Paris* 5 (1930): 215–23 (216).

7 Our hypothetical student could have also been a *collège* pupil: *collèges* were similar institutions to *lycées*: they followed the same syllabi, although they depended for their administration on local authorities, and their teachers only needed a *licence d'enseignement* (teaching degree), while *lycées* were directly managed by the state and their teachers had obtained the *agrégation* (national competitive examination to recruit teachers for the upper forms of the *lycées*); see Célestin Bouglé (ed.), *Encyclopédie française, vol. XV: Education et Instruction* (Paris: Société de gestion de l'Encyclopédie française, 1939), pp. 6–12. Many reforms took place in the period covered in the present book; as a consequence, the rules of access to university varied as well. Although published long ago, Antoine Prost's *Histoire de l'enseignement en France* (Paris: Colin, 1968) is still a good source of information about the French education system up to the 1960s. It contains a handy index of education laws and regulations, arranged in chronological order: pp. 501–511.

8 Cousin's philosophy and its connection with the history of philosophy found a faithful defender in Paul Janet: see Paul Janet, *Victor Cousin et son œuvre* (Paris: Calmann Lévy, 1885). For Janet it was important to present Cousin as the founding father of the history of philosophy in France, and as a consequence dismissed previous examples with the only exception of Joseph Marie de Gérando. More than a century later, Yvon Belaval preferred to write that Cousin 'renovated' the history of philosophy: Yvon Belaval (ed.), *Histoire de la philosophie* (3 vols; Paris: Gallimard, 1973), p. 425. See also Martial Gueroult, *Dianoématique, livre I: L'histoire de l'histoire de la philosophie, 3: en France de Condorcet à nos jours* (Paris: Aubier, 1988), p. 714; and for a comprehensive discussion of the eclectic view of history of

of philosophy disappeared. However, it should be noted that, if the heading 'history of philosophy' was indeed removed, the 'philosophical texts' rubric was still present: teachers were still asked to choose from a list of classics of philosophy, ranging from the works of Xenophon to those of Herbert Spencer. The removal of the history of philosophy from the syllabus has been seen as a weakening of the Cousinian influence on the syllabi.[9] This claim is correct; however, as I shall discuss later in this chapter, I think one should be careful not to take the link between history and philosophy and the Cousinian heritage too much at face-value, especially later in the twentieth century. Some modern commentators rightly emphasize the link that in the nineteenth century existed between Cousin, the spiritualist school and the history of philosophy.[10] That Cousin and Paul Janet played a crucial role in introducing the history of philosophy into the school syllabi is a fact. However, as we shall see in the next chapter, Cousin's philosophy was the target of historians of philosophy's criticism already at the end of the nineteenth century, and more fully at the beginning of the twentieth century. As far as twenty-century historians of philosophy are concerned, to read their work simply as part of the eclectic and spiritualist heritage would be a gross simplification, so extreme, in fact, that it would prevent us from understanding the debates and ideas of the time. The eclectic view of the history of philosophy was not very different from a collection of ideas and texts made into a coherent whole, which was supposed to represent the *perennis philosophia*.[11] This is only one of the ways in which philosophers have conceived of the history of their discipline. On the other hand, many philosophers, including those who had nothing to do with eclecticism, or who openly criticized it, shared the view that philosophy aims at eternal truths, as we shall see in the next chapter. The history of philosophy in France, and indeed elsewhere, has been practised and promoted by scholars with a variety of metaphysical, epistemological and historiographical assumptions.[12]

In the syllabi that followed, the history of philosophy was only reinstated as one of the options: teachers were asked to choose three options from a list of topics and a list of texts (similar to that of the 1902 syllabus); at least one of the options had

---

philosophy see John I. Brooks, *The Eclectic Legacy: Academic Philosophy and the Human Sciences in Nineteenth-Century France* (Newark and London: University of Delaware Press and Associated University Press, 1998).

9 Jean-Louis Fabiani, 'Les programmes et les œuvres: professeurs de philosophie en classe et en ville au tournant du siècle', *Actes de la recherche en sciences sociales* 47–8 (1983): 3–20.

10 See Brooks, *The Eclectic Legacy.*

11 See Gueroult, *Dianoématique, livre I: L'histoire de l'histoire de la philosophie, 3*, p. 739; Fabiani, *Les philosophes de la République*, p. 48.

12 My own early training in philosophy at an Italian *liceo* was in the history of philosophy. Historically, that approach was a legacy of Italian neo-idealism, and the syllabi devised by Giovanni Gentile that still survived in the 1980s. Neo-idealism, which strongly promoted the history of philosophy, had very different assumptions from eclecticism and a very different way of conceiving of the history of philosophy. The history of philosophy thrived then and thrives now in Italy, long surviving the demise of neo-idealism.

to be a text, and all three could be.[13] The list of questions included the history of philosophy along with disciplines that had more recently established themselves, like experimental psychology, pathological psychology, sociology, along with aesthetics (disappeared from the main part of the syllabus), 'science of language' and formal logic. The syllabus changed after the Second World War: in 1960 the questions were listed in two big groups ('knowledge' and 'action') and the list now included the twentieth-century French philosophers Henri Bergson, Alain (Emile Chartier) and Jules Lachelier, and other philosophers, including Marx, Hegel and Nietzsche, who did not appear in previous syllabi.[14]

The fortunes of the history of philosophy at secondary level varied from the end of the nineteenth century to the Second World War. In general, this branch of philosophy did not have a dominant position, but students would have been exposed to either the history of philosophy as such, or a choice of texts by past masters.

The Strength of the History of Philosophy in Higher Education

In higher education, the history of philosophy played a far more important role than it did in *lycées*; indeed, the history of philosophy courses and related examinations formed a crucial part of the philosophy programme at the University of Paris. In the 1920s, the Faculty of Letters offered three history of philosophy examinations (*certificats d'études supérieures*): general history of philosophy, history of ancient philosophy and history of modern philosophy;[15] the other philosophy *certificats* were: general philosophy and logic, ethics and sociology,[16] philosophy of the sciences, aesthetics and science of art, psychology, science of education and sociology.[17] This good offering of history of philosophy courses (good especially considering how vast the intellectual field of 'philosophy' was) was well justified: the 1920 reform established that one of the four *certificats* needed for a *licence*[18] had to be in the history

13 For the 1925 and 1942 syllabi, see François Chatelet, *La philosophie des professeurs* (Paris: Grasset, 1970); see also Emile Bréhier, 'L'histoire de la philosophie dans l'enseignement', *Les Etudes philosophiques*, nouvelle série, 3/4 (1946): 211–17 (211).

14 Syllabi reproduced in Chatelet, *La philosophie des professeurs.*

15 For the sake of simplicity, I translate 'philosophie moderne' with 'modern philosophy', but note that 'philosophie moderne' includes what in English is more usually known as early modern philosophy. As always, boundaries were not given once and for all: Léon Brunschvicg agreed with his colleague Bréhier that modern philosophy started in the eighteenth century, but reported Harald Höffding's view that the modern period started with humanism and the Renaissance: Léon Brunschvicg, *L'esprit européen* (n.p.: La Baconnière, 1947), p. 84.

16 The examination of ethics and sociology in 1920 sanctioned a space for sociology within philosophy: see Victor Karady, 'The Durkheimian Sociology in Academe. A Reconsideration', in Philippe Besnard (ed.), *The Sociological Domain: The Durkheimians and the Founding of French Sociology* (Cambridge: Cambridge University Press, 1983), pp. 71–89 (77).

17 Faculté des Lettres, 'Rapport annuel du Doyen: année scolaire 1925–26', *Annales de l'Université de Paris* 2 (1927): 295–309 (296–9).

18 A *licence* was the first-level university degree, generally made up of four *certificats* (but depending on the subjects the number could range between three and five).

of philosophy. The history of philosophy was also compulsory for teaching degrees (*licences d'enseignements*), which were required in order to gain the *agrégation* and teaching posts in secondary education.[19] Consistently, one of the three written essays of the *agrégation* examination was in the history of philosophy.[20] Starting in the very late 1920s, examinations in new disciplines appeared, notably the history and philosophy of the sciences (as a replacement for the philosophy of the sciences) and ethnology.[21]

While the history of philosophy appeared to shrink in the *lycée* syllabi in the first half of the twentieth century, at university level, and in particular at the Sorbonne, it underwent an expansion, which was consolidated in the inter-war period. Fabiani has provided clear data that reveal the initial crucial leap that the institutional and intellectual profile of history of philosophy took during the first part of the Third Republic (1870–1914): he points out that in that period the largest group of doctoral theses were in the history of philosophy (43 per cent), while the philosophers who held the most important posts had little interest in the history of philosophy, a situation that changed after the First World War.[22] The situation changed indeed: the historians of philosophy who wrote their dissertations in the period studied by Fabiani came to dominate French philosophy in the inter-war years; these included Léon Brunschvicg (doctorate obtained in 1897), Emile Bréhier (1908), Etienne Gilson (1913) and Léon Robin (1908).

The official data at the end of the inter-war period are not easily comparable with Fabiani's data relating to the previous period, because the official publications do not follow the same criteria (i.e., what counts as a history of philosophy topic) and because in the meantime the specializations and branches of philosophy had multiplied. However, it is undeniable that towards the end of the inter-war period, doctoral research in the history of philosophy was very healthy indeed: the 1938 official list of theses deposited at the University of Paris (not yet defended) saw the history of philosophy as the single most popular branch for doctoral dissertations. Out of 421 philosophy dissertations, 161 were in the history of philosophy; 24 in general philosophy and metaphysics; 10 in Logic; 19 in the philosophy and history of the sciences; 17 in ethics, 68 in sociology; 71 in psychology; 26 in pedagogy and 25 in aesthetics.[23] It is worth noting that sociology included a section of 'history of

---

19 Brunschvicg, 'La philosophie dans l'enseignement supérieure à Paris', p. 216, n. 1; International Federation of University Women, *Lexique international des termes universitaires* (Paris, London: Fédération Internationale des femmes diplômées des universités, 1939), p. 243.

20 Bréhier, 'L'histoire de la philosophie dans l'enseignement', p. 211.

21 Cf. Faculté des Lettres, 'Rapport annuel du Doyen: année scolaire 1927–28', *Annales de l'Université de Paris* 4/6 (1929): 445–73 (450–51); Faculté des Lettres, 'Rapport annuel du Doyen: année scolaire 1933–34,' *Annales de l'Université de Paris* 10/5 (1935): 385–404 (392–5).

22 Fabiani, *Les philosophes de la République*, pp. 82–4.

23 History of philosophy theses were divided according to period: antiquity (22 theses); patristics and the Middle Ages (39); early modern period [*époque moderne*] (67); modern period [*époque contemporaine*] (33); the sociology theses were divided in four subcategories:

doctrines', with theses on Comte, Adam Smith, Marx, Rousseau and others. Theses on Rousseau (and on Voltaire and Port Royal) also appeared under ethics.

This large number of philosophy dissertations should not lead us to believe that each year almost 150 students obtained their doctorate, which would be the case if degrees took three years. As doctorate candidates generally were already working, it comes as no surprise that theses could take a long time to complete. Many dissertations still deposited in 1938 had been started six, ten or even fifteen years earlier. In fact, the number of awarded doctorates was relatively small. For instance, the year Jean Wahl obtained his doctorate (1920–21), he was one of 32 candidates to defend their theses in the Paris Faculty of Letters. Of these, six candidates wrote theses in philosophy broadly intended and including psychology, ethnology and sociology. Four out of the six candidates submitted history of philosophy dissertations. As was the rule, Wahl submitted a main dissertation (on the English and German pluralist philosophers) and a complementary one (on Descartes). Gaston Bachelard's theses, supervised by Abel Rey and Brunschvicg, and defended in 1927–28, were in the history and philosophy of science. In that academic year, out of 23 candidates, apart from a thesis on the aesthetic education of the infant, the only other philosophy candidate wrote both his theses on Descartes. The following year (1928–29) Alexandre Koyré was awarded his doctorate with a main thesis on the philosophy of Jacob Boehme; that year virtually all philosophy theses were historical, the others being on Hegel and Taine, on Rousseau and Tolstoy, on Alexandre Herzen, and on Comte.[24]

A simple count of dissertations by no means gives a complete picture of academic intellectual life, but it is an important indicator, especially when, as in this case, it backs up the opinion expressed by one of the most important philosophy professors. Léon Brunschvicg, in an article on philosophy in higher education in France, wrote that an interest in history was a specific trait of philosophy at the Sorbonne, which of course was by far the most important higher-education institution in the country. In general, he judged psychology to be the most important branch of philosophy, and granted the history of philosophy second place.[25] This view obviously reflected what Brunschvicg thought of philosophy in France, but it is also a good, if rough, description of his own interests: intellectual history and the mind. As far as his Sorbonne colleagues were concerned, he believed that it was the history of philosophy in particular that interested them.

How to Become a Leading Professor of Philosophy: Education and Early Career

A large number of important university professors had been educated at prestigious Parisian *lycées*, and the Ecole normale superiéure. The link between a prestigious Parisian *lycée*, the Ecole normale supérieure and the best careers cannot be overstated. Many of the most successful philosophers in this book (though by no means all)

---

archaic societies (5), modern societies (27), history of doctrines (18), social economy and politics (18): *Annales de l'Université de Paris* 13 (1938): 68–87.

24 Albert Guigue, *La Faculté des Lettres de l'Université de Paris depuis sa fondation (17 mars 1808) jusqu'au 1er Janvier 1935* (Paris: Alcan, 1935), pp. 170–72.

25 Brunschvicg, 'La philosophie dans l'enseignement supérieure à Paris'.

followed that route. It has also to be borne in mind that from the early 1880s onwards, a number of Parisian *lycées* started offering courses to prepare to the ENS admission exam, which were popularly called *khâgne*. The *lycée* Louis-le-Grand started to offer this course; the Henri-IV and then the Condorcet and Lakanal soon followed. Although it was possible to prepare in provincial *lycées*, the overwhelming majority of successful candidates came from the Parisian *lycées*, with only the *lycée* in Lyon managing to be a contender.[26] Even before knowing each other as ENS students, many philosophers examined in this book met in one of those *lycées* as fellow pupils, or as teacher and pupil, or were taught by the same teachers in different years. For instance, several of them were *lycée* Condorcet alumni (and *normaliens*), including Henry Bergson (1859–1916), Dominique Parodi (1870–1955), and Léon Brunschvicg (1869–1944). Brunschvicg's schoolmates included not only Marcel Proust, but also Xavier Léon and Elie Halévy, with whom he would collaborate in ventures like the *Revue de métaphysique et de morale* and the Société française de philosophie.[27] Among the Louis-le-Grand *lycée* alumni (and then *normaliens*) discussed in this book there were Victor Delbos (1862–1916), Abel Rey (1873–1940), Emile Bréhier (1876–1952) and Jean Laporte (1886–1948). André Lalande (1867–1963), Célestin Bouglé (1870–1940), who eventually directed the Ecole normale supérieure, and Etienne Gilson (1884–1978) attended the *lycée* Henri-IV. Some of them went on to teach in these same *lycées*, for instance Lévy-Bruhl taught at Louis-le-Grand; Bergson and Delbos at the Louis-le-Grand and Henri-IV; Lalande at the Louis-le-Grand, Condorcet and Henri-IV; Brunschvicg at the Henri-IV and Condorcet.

The connection between having a successful career as a philosophy teacher and attendance of the Ecole normale supérieure is quite clear. Jean-François Sirinelli, in his splendid analysis of the *khâgne* and the Ecole normale supérieure in the inter-war period, has pointed out that the *normaliens* who took the philosophy *agrégation* had almost a monopoly of the higher placements in this competitive examination, superior, for instance, to the results achieved by their fellow *normaliens* in the history and geography *agrégation*. Sirinelli lists a few interesting cases: in 1927, three *normaliens* were first, second and fourth in the philosophy *agrégation*, including Georges Canguilhem (second); in 1929 only Simone de Beauvoir (second) managed to break the dominance of the *normaliens* who took the other four of the five first places (Jean-Paul Sartre was first and Jean Hyppolite third).[28] In addition

---

26 Jean-François Sirinelli, *Génération intellectuelle: khâgneux et normaliens dans l'entre-deux-guerres* (Paris: Presses Universitaire des France, 1994), ch. 3, 'Paris, Lyon et le désert français'.

27 See Christophe Charle, *Les professeurs de la Faculté des Lettres de Paris. Dictionnaire biographique 1909–1939* (Paris: Institut national de Recherche Pédagogique/Editions du CNRS, 1986), pp. 44–5; René Boirel, *Brunschvicg: sa vie, son œuvre, avec un exposé de sa philosophie* (Paris: Presses Universitaires de France, 1964); Gary Gutting, *French Philosophy in the Twentieth Century* (Cambridge: Cambridge University Press, 2001), pp. 40–48 (on Brunschvicg) and ch. 3 (on Bergson); Jean-Louis Fabiani, 'Brunschvicg (Léon)', in Jacques Julliard and Michel Winock (eds), *Dictionnaire des intellectuels français: les personnes, les lieux, les moments* (Paris: Seuil, 1996), pp. 197–8; Christophe Charle, 'Bergson (Henri)', in Julliard and Winock (eds), *Dictionnaire des intellectuels français*, pp. 142–4.

28 Sirinelli, *Génération intellectuelle*, p. 155. Sartre was on his second attempt.

to those mentioned by Sirinelli, many of the philosophers discussed here, who were *normaliens*, performed admirably; for instance Lévy-Bruhl (1879), Brunschvicg (1891), Laporte (1909) and Wahl (1910) were all first in their respective years, Bergson (1881) and Rey (1896) second. Gilson (1907) and Robin (1891) were 'only' sixth.[29]

In the declaration of the convention that established the Ecole normale supérieure in 1794, it was stated that the Ecole would be in Paris 'where citizens from all corners of the Republic who are already educated in the useful sciences will be instructed in the art of teaching by the most skilled of professors in all fields'.[30] In fact, at least in the period covered here, it was preferable to be in Paris already. For a successful career it was advisable to move to Paris as soon as possible, because Paris was where the most important universities, libraries, archives, intellectuals, publishers and institutions were.[31] Even the location of the prestigious educational institutions seems to suggest a link between them: the Ecole normale supérieure was (and is) on the Parisian rue d'Ulm,[32] very close to the *lycées* Henri-IV and Louis-le Grand: all three institutions are nestled around the Pantheon, where French 'great men' are buried, and are within a short stroll of the Sorbonne and the Collège de France.

It was not unusual that Delbos, Lévy-Bruhl, Brunschvicg and Lalande taught in *lycées*: in the period between 1909 and 1939, the first teaching post of 75 per cent of Sorbonne professors had been in a *lycée*.[33] If, as was often the case, they started their career in a provincial lycée, as a rule their aspiration would be to move to Paris.[34] Failing to reach Paris as a *lycée* teacher, the most common route was through a provincial university (the first appointment of 17.7 per cent of Sorbonne professors

29 See Charle, *Les professeurs de la Faculté des Lettres de Paris*; Christophe Charle and Eva Telkès, *Les professeurs du Collège de France. Dictionnaire biographique (1901–1939)* (Paris: CNRS, 1988); Guigue, *La Faculté des Lettres de l'Université de Paris*. There were of course exceptions; for instance Albert Rivaud (1876–1956), who was first at the philosophy *agrégation* in 1900, was not a *normalien*.

30 Website of the Ecole Normale Supérieure: http://www.ens.fr/ecole/presentation.php (accessed 2 April 2007).

31 As Charle has put it, for a professor of letters not to go to Paris meant to place oneself at the margins of the scholarly world: Christophe Charle, *La république des universitaires, 1870–1940* (Paris: Editions de Seuil, 1994), p. 191.

32 I am referring to 'the' Ecole normale, which was attended by young men ('Ecole normale supérieure de la rue d'Ulm'). The Ecole normale for young women ('l'Ecole normale supérieure de jeunes filles') sited at Sevrès, was founded in 1881, 107 years after the former, and merged with it in 1985; see the Ecole normale supérieure website. For the attempt to create a female professor during the Third Republic, see Jo Burr Margadant, *Madame le professeur. Women Educators in the Third Republic* (Princeton, NJ: Princeton University Press, 1990).

33 See Charle, *La république des universitaires, 1870–1940*, p. 82.

34 For example, Brunschvicg started teaching in 1891 in provincial *lycées* (in Lorient, Tours and Rouen), then he was appointed to the Parisian *lycée* Condorcet first and then to the Henri-IV; in 1909 he obtained his first university post, in Paris. Henri Bergson began his career in 1881 at the *lycées* of Angers and Clermond-Ferrand, then moved to the Parisian *lycée* Louis-le-Grand, the *collège* Rollin and the *lycée* Henri-IV. His first post in higher education came in 1898, at the Ecole normale supérieure, and in 1900 he was elected to a chair at the Collège de France.

had been at a provincial university). An example of the latter route is the career of Abel Rey, a provincial himself, who attended the *lycée* in Marseilles and then the Parisian *lycée* Louis-le-Grand. After his degree at the University of Paris and his *agrégation*, he taught in provincial *lycées* between 1898 and 1908. He started his university career in a provincial university (Dijon), where he became professor in 1911, but in 1919 he moved to the University of Paris when he was first appointed *maître de conferences*,[35] and after six years, professor. Similarly, Emile Bréhier, who started his teaching career in provincial *lycées* (1900–1909), took his first university post at the University of Rennes, moving to Bordeaux and eventually, in 1919, to the University of Paris, where he started as *maître de conférences* in the history of philosophy, although he had been a professor in Bordeaux. He eventually became one of the most important and influential Sorbonne professors, and one of the most eminent historians of philosophy.

Scholars who became the most important academics often taught generations of pupils at secondary level. Moreover, some philosophers who published and took part in the philosophical life in the capital spent their whole life teaching in a *lycée*. A good example among historians of philosophy is André Cresson, whose friendship with Léon Brunschvicg began at the *lycée* Condorcet where they were both pupils.[36] He taught in all the great Parisian *lycées*, was the author of a remarkable number of introductions to philosophers published by Alcan and Presses Universitaires de France, and of numerous articles in the *Revue de métaphysique et de morale*; he also regularly attended the meeting of the Société française de philosophie, and was a member of the Société des amis de l'histoire des sciences.[37] The most famous lycée and *khâgne* teacher, however, must be Emile Chartier, known as Alain.[38] He taught for 34 years at the *lycée* Henri-IV, where philosophy in the *khâgne* had been taught before him by Henri Bergson, Victor Delbos and Léon Brunschvicg.[39] Alain's students formed a fairly recognizable group, influenced not only philosophically, but

35 The *maîtres de conférences* were roughly equivalent to lecturers in British universities. For the main documents that established the duties of the *maîtres de conférences*, see Guigue, *La Faculté des Lettres de l'Université de Paris*, pp. 32–7.

36 See André Cresson, 'Lettre sur Léon Brunschvicg', *Revue de métaphysique et de morale* 50 (1945): 5–7.

37 See Fabiani, *Les philosophes de la République*, p. 32, n. 29; Francis Ambrière, 'Société des amis de l'histoire des sciences', *Thalès* 3 (1936): 249–61 (255); Hélène Metzger expressed a positive judgement of his work, writing to Sarton that his two volumes on the history of thought from the seventeenth century to the present went well beyond a mere popularization of philosophy (Hélène Metzger, Letter to Sarton of 28 February 1927, Houghton Library, bMS Am 1803 [1032]). She was probably referring to his work in two volumes on French philosophy: André Cresson, *Les courants de la pensée philosophique française* (Paris: Colin, 1927). Many of his books on individual philosophers had several editions, and their topics ranged from Aristotle and Marcus Aurelius to Rousseau, Kant, Nietzsche, Bergson and many others.

38 See André Sernin, *Alain: un sage dans la cité* (Paris: Robert Laffont, 1985) or, for a short biography, André Sernin, 'Alain [Emile Chartier]', in Julliard and Winock (eds), *Dictionnaire des intellectuels français*, pp. 43–4.

39 Sirinelli, *Génération intellectuelle*, p. 87.

also ethically and politically, by their pacifist, Dreyfusard, anticlerical and radical teacher.[40] Among the many students who stayed in contact with Alain after their *lycée* years, and collaborated with their teacher's magazine, *Libres Propos*, there were Simone Weil and Georges Canguilhem. One wonders whether Brunschvicg, when saying that university students were 'no novices' to philosophy, gave a little private sigh, thinking of those among his students who had been trained by Alain. In her biography of Weil, Simone Pétrement recalls the hostility that Alain's former students, including herself, displayed against Brunschvicg, their lecturer at the Ecole normale supérieure.[41] For his part, Brunschvicg reportedly found Alain's legacy to his students quite awkward: he lamented that they, when asked to comment on a philosophical text, would first declare that one should bracket all interpretations and go back to the text in its purity, and would then proceed to repeat Alain's interpretation.[42] University professors had to deal with their students' pre-university training and they often had experience of imparting this training.

Professors of the History of Philosophy and their Social Backgrounds

Academics at the University of Paris and the Collège de France had been very successful at the various stages of their education and career. In order to understand the field of the history of philosophy, a closer look at the professors teaching the subject at the Sorbonne and Collège de France is in order. This of course does not give us a complete picture of scholars or even only of academics carrying out research in the history of philosophy. It goes without saying that other scholars published works in the history of philosophy, such as the already mentioned Cresson, and others, some of whom we shall meet in the following chapters. However, it will help us understand whether the professors of history of philosophy had power within academic philosophy, and as a consequence whether they were important in setting research agendas and promoting younger academics. It will also be interesting to find out the background, or the social and cultural capital – in the terminology of Pierre Bourdieu – of the scholars appointed to the most prestigious chairs in the history of philosophy in the first half of the century. The interest of this analysis is twofold: their social background will shed light on their career opportunities and choices and consequent positions in academia. More ambitiously, I shall also discuss – later in the book – whether there was any link between their social and religious background and their ideas about the mind and its history.

My temporary focus on individuals is not aimed at refuting Terry Clark's theory of 'clusters', according to which:

> From the standpoint of intellectual organization and innovation the most important unit was not the individual incumbent of a chair, but an informal grouping that might be termed a 'cluster'. The cluster was an association of perhaps a dozen persons who

40 See ibid., ch. 13: 'Les élèves de Alain'.

41 Simone Pétrement, *La vie de Simone Weil*, vol. 1: *1909–1934* (Paris: Fayard, 1972), pp. 128–9.

42 Raymond Aron, *Mémoires* (Paris: Robert Laffont, 2003 [1983]), pp. 43ff.

> shared a minimal core of beliefs about their work and who were prepared to collaborate to advance research and instruction in a given area.[43]

Even less do I want to offer a criticism of his critics, such as Fabiani, who thinks that Clark's theory of clusters is superfluous in order to understand the logic of philosophical schools and groups.[44] In fact, it is not my aim to put forward any general theory about French academia, and not even about all philosophers. My focus in this chapter is the history of philosophy and I shall examine the activities and ideas of its practitioners. In the following chapters, I shall also consider other scholars, who aimed to study the mind in a historical and philosophical manner. I shall draw conclusions that will only apply to them, and might well not apply to other scholars, disciplines or subdisciplines. My first step here is to focus on individuals, but in fact I shall consider groups and institutions, and focus on discussions, notably those at the Société française de philosophie and the Centre de synthèse.

Between 1900 and 1939, only one historian of philosophy was elected at the Collège de France:[45] Etienne Gilson, who, in 1932, obtained the chair in History of Philosophy in the Middle Ages. He had previously been *maître de conférences* at the Sorbonne and director of studies at the Ecole pratique des hautes études (both from 1921), after teaching in provincial *lycées* and at the Universities of Lille and Strasbourg. The duties of a professor at this highly prestigious institution did not include any formal teaching, but only public lectures.[46] At the Paris Faculty of Letters, the history of philosophy had much more space. Between 1908 (when the first appointment in the history of philosophy of the new century was made) and 1939 nine professors of history of philosophy appointed: Lévy-Bruhl, Brunschvicg, Rodier, Bréhier, Robin, Delbos, Rivaud, Laporte and Milhaud. Their teaching was linked to five chairs:

1. Chair of History of Modern Philosophy. In 1908, Lucien Lévy-Bruhl obtained this prestigious chair, which had been left vacant by Emile Boutroux, who in turn had replaced Paul Janet. It had been created in 1888 as Philosophy and Opinions of Philosophers, a name that changed several times (History of Philosophy, History of Modern Philosophy, then back to History of Philosophy, and once again History of Modern Philosophy).[47] Lévy-Bruhl was charged with a course of History of Modern Philosophy in 1902, and appointed *adjoint*

43 Terry N. Clark, *Prophets and Patrons: The French University System and the Emergence of the Social Sciences* (Cambridge, MA: Harvard University Press, 1973), p. 67.

44 Fabiani, *Les philosophes de la République*, p. 77.

45 On the whole, 64 appointments were made in the humanities and social sciences: see Charle and Telkès, *Les professeurs du Collège de France*, p. 16.

46 On the Collège de France, see Christophe Charle, 'Le Collège de France', in Pierre Nora (ed.), *Les Lieux de mémoire, II: La Nation* (Paris: Gallimard, 1986), vol. 3, pp. 389–424.

47 For the creations of chairs and their changes of name see Guigue, *La Faculté des Lettres de l'Université de Paris*, pp. 14–32.

professor in 1904.[48] He retired in 1926, and in 1927 Léon Brunschvicg, who had been *maître de conferences* at the Sorbonne since 1909, was nominated to this chair, which he held until 1940.

2. History of Ancient Philosophy. Louis Rodier was the first professor to be appointed to the chair of History of Ancient Philosophy in the twentieth century; he died in 1913, only four years after his appointment; the chair was suppressed and then re-instated for Léon Robin (1924–1936). Robin had been teaching in the Faculty of Letters since 1913.
3. Philosophy and History of Philosophy. This chair was created in 1913 for Victor Delbos, who had been professor of Philosophy and Psychology from 1909 to 1913. Delbos died in 1916, and his chair was suppressed, to be then recreated in 1933 for Albert Rivaud, who had been professor without chair since 1929. Rivaud officially retired in 1945, but he had been suspended in 1944 for his collaboration with the occupying army. Emile Bréhier is normally named professor of Philosophy and History of Philosophy from 1930 to 1946.[49] Strictly speaking, in 1930 Bréhier was nominated '*professeur titulaire à titre personnel*' in Philosophy and History of Philosophy. This, however, did not make any real difference, for these professors had the same status as those who held a chair.[50] He had been nominated professor without chair in 1923, four years after obtaining his post of *maître de conférences* at the Sorbonne. He retired in 1946, and was made Professor Emeritus.[51]
4. History of Medieval and Modern Philosophy. Jean Laporte, who had been appointed *maître de conferences* of History of Medieval and Modern Philosophy in 1932, became professor 'without chair' in 1933. Eventually, in 1937, he obtained the chair of History of Medieval and Modern Philosophy. He had already lectured at the Sorbonne between 1925 and 1931, replacing Bréhier, Lalande, Robin and Rey while they were teaching abroad.
5. History of Philosophy in its Relationship with the Sciences. Gaston Milhaud was appointed in 1909 to this chair; he died on active service in 1918.

Who were these Sorbonne and Collège de France historians of philosophy? What social and cultural background did they have? From the point of view of their social capital, the nine professors appointed in the Faculty of Letters and the one appointed at the Collège de France between 1908 and 1939, formed a relatively heterogeneous group, bearing in mind that at this point in history, the

48 A '*professeur adjoint*' was a *maître de conférences* chosen by the council of their faculty to enjoy the privileges of the '*professeurs titulaires*', that is, professors who held a chair: International Federation of University Women, *Lexique international des termes universitaires*, p. 232.

49 For instance in the entry on him in Charle, *Les professeurs de la Faculté des Lettres de Paris*.

50 See Guigue, *La Faculté des Lettres de l'Université de Paris*, p. 30.

51 Ibid., p. 258.

working and peasant classes were nearly absent from such posts.[52] Only three professors had a background in the intellectual professions: Bréhier, Rodier and Delbos. Bréhier's father was deputy head in a *lycée* (grammar *agrégé*) and one of his uncles was a university professor and member of the Institut de France. The background of the other, Rodier, was not as close to the teaching professions as Bréhier: his father was an assistant clerk and librarian of the Court of Appeal. Only for his role as librarian can we put him in the same broad category as Bréhier. Victor Delbos' father was a notary clerk and his mother a librarian. At the top of the social ladder we find Albert Rivaud and Jean Laporte. The latter was the son of a china industrialist in Limonges. Rivaud was the only professor appointed at the Faculty of Letters in this period who was the son a prefect: in fact, only 4.7 per cent of them were children of senior civil servants. Milhaud's and Robin's fathers were merchants [*négociants*], while Gilson's father was a hosiery salesman.

Lévy-Bruhl and Brunschvicg had the weakest social capital among these history of philosophy professors. Like Gaston Milhaud they were Jewish, but their social background was rather different from his. Lévy-Bruhl's father was a sales representative (seven other sales representative's children were appointed in the Faculty of Letters between 1909 and 1939) and Brunschvicg's a manufacturer of furnishing trimmings (the only one in this period). Like Milhaud, who moved from the University of Montpellier to take up a chair in Paris in 1909, Lévy-Bruhl and Brunschvicg, who were appointed *maître de conferénces* in the Faculty of Letters in Paris in 1899 and 1909 respectively, were among the first people of Jewish origin to pursue an academic career. As a consequence, it comes as no surprise that their fathers were not academics. However, their origin was humbler than that of most of their colleagues.

52 I extract these data from Charle, *Les professeurs de la Faculté des Lettres de Paris. Dictionnaire biographique 1909–1939*. The attribution of a certain occupation to a certain social class in the France of the time comes from this work.

Table 1 Social and religious background of the professors appointed to History of Philosophy chairs at the Sorbonne and the Collège de France, 1900–1939.

Name	Chair (year of appointment)*	Father's profession	Family of origin's religion
Lévy-Bruhl	History of Modern Philosophy (1908)	Sales representative	Jewish
Brunschvicg	History of Modern Philosophy (1927)	Manufacturer of furnishing trimmings	Jewish
Rodier	History of Ancient Philosophy (1909)	Assistant clerk and librarian, Court of Appeal	Catholic
Robin	History of Ancient Philosophy (1924)	Merchant	Unknown
Delbos	Philosophy and History of Philosophy (1913)	Notary clerk (mother: librarian)	Catholic
Bréhier	Professeur à titre personnel, Philosophy and History of Philosophy (1930)	*Lycée* deputy head	Catholic
Rivaud	Philosophy and History of Philosophy (1933)	Prefect	Catholic
Laporte	History of Medieval and Modern Philosophy (1937)	China industrialist in Limonges	Catholic
Milhaud	History of Philosophy in its Relationship with the Sciences (1909)	Merchant	Jewish
Gilson	History of Philosophy of the Middle Ages, Collège de France (1932)	Hosiery tradesman	Catholic

Sources: Gigue, *La Faculté des Lettres de l'Université de Paris*; Charle, *Les professeurs de la Faculté des Lettres de Paris. Dictionnaire biographique 1909–1939*; Charle and Telkès, *Les professeurs du Collège de France*.

* The year of appointment to a named chair does not correspond to the year of appointment at the Sorbonne, which was normally many years earlier (for instance, Brunschvicg had his first appointment there 17 years before obtaining his chair). For details, see the list of chairs in the text. When the institution is not stated, it is the Sorbonne.

The importance of inherited social capital for access to higher education has been a topic of intense research and discussion for decades. Jean-François Sirinelli thinks that Pierre Bourdieu and Jean-Claude Passeron's thesis that French higher education is an instrument of social reproduction, while valid for the period they examined (1960s), cannot be extended to previous periods. He has argued that it cannot be applied to the period he has examined – the inter-war years – in which he has noticed a wider social spectrum among *normaliens*.[53] Without aiming to put forward a general theory of social reproduction in France, it is fitting to notice that among philosophers, and academics in general, there was indeed some social mobility in the first decades of the twentieth century. The rapid expansion of tertiary education must have removed some of the obstacles for entering it. Among the cohort of Parisian historians of philosophy, the percentage of individuals whose fathers were not in an intellectual profession is high, relatively speaking. This information about their background prevents us from applying to them the general statement that university professors in the Third Republic 'were typically born into intellectual families', and as a consequence were 'less inclined to identify with the conservative values of the wealthy bourgeois class'.[54] Seven out of these ten professors of history of philosophy were not born into intellectual families (and in reality the percentage of professors from intellectual families appointed at the Paris Faculty of Letters in 1909–39 was 'only' 35.8 per cent).[55] Their social background was rather diverse, ranging from wealthy industrialist (Laporte) and senior civil servant (Rivaud) to small trader (Brunschvicg) and modest sales representative (Lévy-Bruhl). The Jewish background of three of them is an important factor, including in their relationship with the republican institutions. They favoured the secular character of the republic, and may have felt that they had reached their positions thanks to a meritocratic system, without the advantage of social connections which their peers from 'mainstream' intellectual families had enjoyed.

However, economic capital made a great difference even within this particular social group. For one, it could not fail to play a role in career choices. A clear example is that of Brunschvicg and Xavier Léon. Much united them, but Brunschvicg could not have made the choices that the latter made. Léon's decision to focus on the editing of a major philosophy journal, and on the organization of international conferences and a debating society would have not been feasible for Brunschvicg, for he lacked economic capital. The latter needed a salaried job; he gained all the academic qualifications to which a student of philosophy could aspire, and pursued a brilliant career in the educational institutions of the Republic. While Léon started his journal, Brunschvicg taught in a provincial lycée. Brunschvicg gained his social position through formal education and qualifications, and it was in the same Republican institutions that gave him his qualifications that he pursued his intellectual career and reached his prominent position. His social connections had been gained thanks to the prestigious schools that he attended, rather than his family.

Personal Strategies and Institutional Success

Lévy-Bruhl and Brunschvicg, who, unlike Milhaud, were both born in Paris (which was a social advantage), followed the most secure strategies to 'correct' their

53 Sirinelli, *Génération intellectuelle*, pp. 167–8.

54 Gutting, *French Philosophy in the Twentieth Century*, p. 7.

55 Charle, *Les professeurs de la Faculté des Lettres de Paris*, p. 3.

unorthodox background.[56] One was a good marriage and the other a brilliant education. The marriage patterns of French intellectuals of Jewish origin has been studied by Christopher Charle, who has taken as his model Emile Durkheim's 'beau marriage' to the wealthy Julie Dreyfus. Relatively poor Jewish men with intellectual ambitions would often marry into wealthy Jewish families.[57] Our historians of philosophy did precisely that: Lévy-Bruhl, who was born Lucien Lévy, married Alice Louise Bruhl, the daughter of a wealthy jewel merchant, and so acquired the second part of his surname.[58] Léon Brunschvicg married Cécile Kahn, who too was the daughter of a wealthy merchant. They married in 1899, when Brunschvicg had already gained all his qualifications and had secured a permanent job in education, and presumably already knew of his transfer from the *lycée* in Rouen to the Condorcet *lycée* in Paris, where he started teaching in 1900. She pursued a political career as a feminist and as a member of the Radical Party, eventually becoming the first woman to enter the French cabinet, when in 1936 she was appointed junior minister for education in the Léon Blum government.

Although acquired economic capital was important, acquired cultural capital was a necessary condition to secure a Sorbonne post. Lévy-Bruhl's and Brunschvicg's education was nothing less than exemplary. Both attended a Parisian *lycée*; at the prestigious *lycée* Condorcet, Brunschvicg not only received a good education, but also established a network of friends that was going to be very significant for his professional life. His excellence in his secondary studies was crowned by his double success at the *concours général* (in French and philosophy), a competitive examination taken, solely for reasons of prestige, by the best pupils of two final years of secondary education. He thus joined the ranks of the winners of the *concours*, who included Turgot, Antoine Lavoisier, Louis Pasteur and Victor Hugo, and, closer to him in time, the philosophers Emile Boutroux, Alain and Henri Bergson, and the politicians Jean Jaurès and Léon Blum.[59]

Brunschvicg and Lévy-Bruhl continued their education in the most prestigious institutions and with remarkable results: they were both *normaliens*, and both were ranked first at their respective *agrégation* exams. Although they taught for 16 and 18 years in *lycées*, respectively, before obtaining their first posts in higher education, they both created a secure power base in the university. Their influence extended to the two most important philosophy journals. In 1916, Lucien Lévy-Bruhl took up the editorship of the *Revue philosophique*, which he kept until his death in 1939; in 1940 Emile Bréhier and Paul Masson-Oursel replaced him. Brunschvicg was from the very beginning in the inner circle of the *Revue de métaphysique et de morale*. He also served as chair of the jury of the *agrégation* examination, and became a member

56 Their families were, however, from Alsace (Brunschvicg) and Metz (Lévy-Bruhl).

57 Charle argues that the class boundaries were weaker among Jews than Catholics; Christophe Charle, 'Le beau marriage d'Emile Durkheim', *Actes de la recherche en sciences sociales* 55 (1984): 45–9.

58 He also acquired a niece, Hélène Metzger, née Bruhl, who in turn married a poor intellectual, Paul Metzger. I shall discuss Metzger's work in Chapter 5.

59 See the website of the Association des lauréats du Concours général: http://concoursgeneral.free.fr/ (accessed 18 April 2007), in particular the page 'Palmarès'. In Brunschvicg's time this competition was limited to Parisian schools.

of the Académie des sciences morales et politiques (Institut de France) in 1919,[60] at fifty years of age, and served as its president in 1931.[61]

Brunschvicg's and Lévy-Bruhl's respective strategies were remarkably successful; from disadvantageous social and religious backgrounds, they rose to be important, well-respected and powerful academics. Within the group of historians of philosophy, the capital of university power possessed by Lévy-Bruhl and Brunschvicg arguably was equalled only by Bréhier, who however came from an intellectual family, including a maternal uncle who was a member of the Institut de France. Bréhier founded the *Revue d'histoire de la philosophie*, and then co-edited the *Revue philosophique*; he also directed the *Nouvelle encyclopédie philosophique*. He too was elected member of the Académie des sciences politiques and philosophiques, but in 1944, two years after his retirement, as a crowning of his career. He also served in the Comité consultatif de l'instruction superiéure, an advisory body to the Ministry of Education. By looking at Brunschvicg's and Bréhier's activities and honours, of which I have only listed a few, it is not very difficult to believe Pierre Bourdieu when he claimed that their influence was still felt in the 1970s, as the 'most prestigious careers' of philosophers in that period depended on having as a thesis supervisor a Sorbonne professor who thirty years earlier was attached to either of them.[62]

Both of them also enjoyed international success, which is an important indicator of scientific power and prestige. Just to cite a few honours, Bréhier was member of Belgium's Académie royale des sciences, des lettres et des beaux-arts, and of the Accademia nazionale dei Lincei; Brunschvicg was member of the Accademia dei Lincei as well, of the Royal Academy of Denmark, and the Accademia reale di scienze morali e politiche of Naples. He also received an honorary degree from Durham University.[63] In this group, Robin too enjoyed international recognition: he was visiting professor at the University of Pennsylvania and at Harvard (1927–28) and exchange professor at Liège, where he was also awarded an honorary degree.

However, arguably nobody among these historians of philosophy enjoyed as much international recognition as Etienne Gilson. He received honorary degrees from 15 Universities including Oxford, Harvard, Milan, Montréal and Columbia, and was elected member of 20 foreign academies, including the British Academy, the Accademia dei Lincei, the American Academy of Arts and Sciences, the Royal Society of Canada and Real Academia española. He was also visiting professor at Harvard and Liège, and delivered the Glifford lectures at the University of Aberdeen

60 The Institut de France includes five academies: Académie française (founded 1635); Académie des inscriptions et belles-lettres (1663); Académie des sciences (1666); Académie des beaux-arts (created in 1816 by merging the Académie de peinture et de sculpture [1648], the Académie de musique [1669] and the Académie d'architecture [1671]); and the Académie des sciences morales et politiques (created in 1795, closed down in 1803 and re-created in 1832). Institut de France's website http://www.institut-de-france.fr/institut/index.html (accessed 20 July 2006).

61 Membership of the Institut de France conferred unparalleled prestige that reinforced academic power. The mention of Institut membership followed scholars' names in their books' title pages and in reports of official functions.

62 Pierre Bourdieu, *Homo Academicus* (trans. Peter Collier; Cambridge: Polity Press, 1988 [1984]), p. 93.

63 See Charle, *Les professeurs de la Faculté des Lettres de Paris*, pp. 41, 45.

in 1930 and 1931. In France, Gilson was at the Collège de France, a prestigious post that afforded great freedom but meant that he did not regularly teach students who were going to be teachers. He was nonetheless part of the teaching establishment: for instance he was member of the philosophy *agrégation* panel (1919–21), and of the examination panel at the Ecole normale supérieure (1930). He was also a public intellectual: just to mention two of his many activities, he regularly wrote in the newspapers *Le Monde*, and in 1945 was part of the French delegation to the San Francisco conference, where the charter of the United Nations was written.

Just as Brunschvicg identified with the French republican institutions and their ethos, Gilson's career was informed by his Catholic faith and his alliance with the Vatican. His Catholicism was very much part of his philosophy, and he defended the possibility of a Christian philosophy against Brunschvicg who denied it.[64] Brunschvicg, Bréhier, Lévy-Bruhl, Milhaud and Rodier were secularists.[65] Their position was much more in tune with the values of the Third Republic than Gilson's, and the mark that at least some of them left on French academia was more durable. Gilson's activities were often linked to a powerful international organization: the Catholic Church. In particular, in 1929 he founded the Pontifical Institute of Medieval Studies in Toronto, where he lectured for one term a year (apart from the war years); in 1951, he retired from the Collège de France and became full-time professor at Toronto. Despite his different trajectory, one should not be tempted to regard Gilson as separated from the secular historians of philosophy. He regarded himself as having been 'profoundly influenced' by his teacher Lévy-Bruhl, whom he saw as the embodiment of the 'great professor of history of philosophy'. However, he immediately added that he did not mean to imply that Lévy-Bruhl would have approved of his way of writing this history, but only to acknowledge his 'immense debt' to his teacher.[66]

Delbos and Laporte were also Catholic, but their religion did not play as central a role in their philosophy and academic life as it did for Gilson. The most extreme position was that of Rivaud, who was a conservative Catholic, with positions close to the monarchist and anti-Dreyfusard Charles Maurras, the *idéologue* of the extreme-right movement Action française. As the child of a prefect, Rivaud was born within the institutions, so to speak, while Brunschvicg had gained his place in them. Those institutions did not protect Brunschvicg during the Second World War: he had to go into hiding, where he died. Rivaud, who was close to Marshall Pétain, on his part, even served briefly in the Vichy government as minister of education in 1940, only to be removed after German pressure, probably for his nationalistic and anti-German views.

Lévy-Bruhl and Brunschvicg had started as the most marginal, for they combined Jewish origins with relatively humble backgrounds. However, they were socially mobile, and occupied posts of great importance in French academia. Although their political convictions were not the same, they shared the faith in the French

64 I shall discuss their exchanges on Christian philosophy in Chapter 3.

65 According to Charle, Robin's religion is 'unknown'.

66 Quoted in Gueroult, *Dianoématique, livre I: L'histoire de l'histoire de la philosophie, 3*, p. 1031, n. 1. This quotation is from a letter that Gilson wrote to Gueroult in 1949.

'republican ideal',[67] and a belief in social progress to be achieved within the French republican institutions.

Social and religious background, along with political and personal choices could not fail to have an impact on the careers of these philosophers. They also played a role in their views of the institutions in which they worked, and on their commitment to them, as well as their commitment to society. What they, and many others whom I shall discuss in the following chapters, had in common was the study of philosophy and of its history. However, their views of their activity did not always coincide. After investigating the historians of philosophy's diverse social, religious and political backgrounds, I shall now turn to their views of the methods and aims of the history of philosophy.

67 Lévy-Bruhl presented his view of the republican ideal in Lucien Lévy-Bruhl, 'L'idéal républicain', *Revue de Paris* 31 (1924): 805–22. His article was republished as a pamphlet by the Editions du progrès civique, see Jean Cazeneuve, *Lucien Lévy-Bruhl: sa vie, son œuvre* (Paris: Presses Universitaires de France, 1963), p. 16.

Chapter 2

The History of Philosophy in the First Decades of the Twentieth Century: Theory and Objectives

Introduction

The history of philosophy played an important role at the Sorbonne, both in terms of chairs and research students and according to the views of philosophers like Brunschvicg; it also played a central role in the research output, as seen in the previous chapter. But what was 'the history of philosophy'? What were its methods and aims? Was its goal the philological erudition exemplified in commentaries, or the engagement with the great problems with which philosophers have grappled across centuries? Was it the faithful rendition of a past scholar's work, or its free interpretation, aimed at creating new philosophical ideas? For a junior scholar of the time, the best way to find out the state of contemporary scholarly discussion about these issues was to read the most important philosophical journals, as well as attending the most prestigious seminars and international conferences. The two great philosophical journals in the first half of the twentieth century were the *Revue philosophique* and the *Revue de métaphysique et de morale*: the former originated from within the Sorbonne, whereas the founder and editor of the latter was not an academic. The most important place for 'synchronous' discussions was the Société française de philosophie, which was an off-shoot of the *Revue de métaphysique et de morale* (*RMM*), and as such, not officially linked to the university. The *RMM* group also created that other crucial place for philosophical discussion: the international congresses of philosophy which took place from 1900 onwards around Europe.

It is in these real or virtual places – journals, Société française de philosophie and international philosophy conferences – that crucial debates about the history of philosophy took place. I shall consider a few exemplary contributions in each medium, the journals first, the Société in the following section, and lastly Bergson's landmark paper at the 1911 International Congress of Philosophy. By travelling through pages and places, we will find out not only what the methods and aims of the history of philosophy were at the time, but also the assumptions that supported what I shall call 'mainstream' conceptions of the history of philosophy. These conceptions were by no means homogenous, but nevertheless, to different degrees, they shared certain fundamental beliefs.

First of all, I shall discuss the opposition of most historians of philosophy to Hegel's, Comte's and Cousin's systems, for being *a priori* and unsupported by empirical evidence, and for undermining the uniqueness and originality of individual philosophical works. In fact, many historians of philosophy considered their direct and creative engagement with past texts as the more authentically philosophical part of their activity. The possibility of this direct engagement was based on the belief that ultimately philosophical texts are timeless, and that the ideas expressed in them can still play an active role in modern philosophy. For most 'mainstream' historians of philosophy, this special status attributed to philosophy sharply distinguished it from science: just as the history of science for them exhibited progress, the history of philosophy did not. I shall also show that Bergson's paper lent theoretical support to a large part of the historians of philosophy's practices, although it fell short of supporting one important role that they attributed to themselves: to preserve a heritage and a tradition. These ideas supporting the epistemology and historiography of the history of philosophy were, however, often challenged, not only by historians and social scientists, but even by historians of philosophy themselves: first of all Lévy-Bruhl and Brunschvicg. I shall discuss their challenges in detail in the following chapter, here I shall present some theoretical points of contentions as they emerged in scholarly discussions. But first, I shall offer some insight in the history and nature of these very special virtual and real spaces.

Academic Journals and the Historiography of Philosophy

The Revue philosophique *and the* Revue de métaphysique et de morale

At the beginning of the twentieth century, the *Revue de métaphysique et de morale* and the *Revue philosophique* were already well-established.[1] The *Revue philosophique*, founded by Theodule Ribot in 1876, can be seen as the first French philosophical academic journal, followed by *Revue de métaphysique et morale* in 1893. It is all too easy to regard these two journals as the output of two different factions in the philosophical world, considering Ribot's interests in experimental psychology and his advocacy of the new social sciences, and the anti-positivist convictions of the founder and editor of the *Revue de métaphysique et morale*, Xavier Léon, along with his defence of philosophy as an autonomous discipline. Even the title of the second journal appears polemical: metaphysics was precisely what for positivists and the advocates of the social sciences could not or should not be done. Indeed, traditionally these two journals have been seen as having opposing aims and practices. In a recent article, Daniela Barberis, while warning against reading too rigid an opposition between these two journals, has nevertheless presented the *Revue philosophique* and the *RMM* as representing 'opposing positions in the philosophical spectrum' and the *RMM* as belonging to the 'spiritualist' camp. She has added that the *RMM*'s inner circle's 'emphasis on the importance of the mind and their hostility to positivism made them ideal heirs for the old philosophical guard'.[2]

---

1 Before the foundation of the *Revue philosophique*, philosophers generally published in general journals, like the *Revue des deux mondes*. Journals that were the output of particular schools also existed, such as *Philosophie positive* and *Critique philosophique*.

2 Daniela S. Barberis, 'Moral Education for the Elite of Democracy: The Classe de Philosophie between Sociology and Philosophy', *Journal of the History of the Behavioral*

Barberis is not wrong in pointing out the differences between the two journals. However, I think that not only some distinctions are in order, but also that a representation of *RMM* as simply reproducing 'old' philosophy is not accurate. Indeed, a focus on the mind in France was going to give impulse to new approaches, which ranged from new studies of the history of science, such as Koyré's and Metzger's, to historical epistemology, as I shall discuss in the last chapters of this book. Even the choice of title, *Revue de métaphysique et de morale*, was not completely straightforward. The title was closely related to the title of Frédéric Rauh's thesis *Fondement métaphysique de la morale*.[3] Rauh was closely involved in the foundation and the running of the *RMM*, but it was he who proposed to Léon the title *Revue de questions philosophiques* rather than *Revue de métaphysique et de morale*, in order to avoid misunderstandings as to the content of the journal.[4] Maybe Rauh already had doubts about metaphysics, doubts that made him turn to psychology a few years later. More to the point, metaphysics was not even embraced by all members of the *RMM* 'inner circle', whose philosophical approaches exhibited important differences. Brunschvicg, who would become the most important philosopher in the group, certainly did not believe that metaphysics could or should be practised. It would therefore be misleading to assume that the *RMM* was the site of old-fashioned metaphysicians, who intended to revive their discipline after the attacks of positivism. As Christophe Prochasson has put it, the *RMM* 'was not reactionary', and did not propose 'any return to forgotten old ideas'.[5]

When discussing the relationship between these two major journals, different perspectives may lead to rather different conclusions. Some scholars have focused on the intentions of their founders, with a particular focus on the very beginning of a journal's history (like Barberis), while others have focused on the actual output of the journal over a longer period of time. It is this second focus that interests me here, for the role played by the *RMM* can be, and should be, separated from Léon's initial intentions and policies, which are beside the scope and the chronological span of this book. Dominique Merllié has adopted the latter perspective, and I think that he has conclusively proved that a careful analysis of the two journals shows that the *RMM* cannot be seen as an 'idealist war machine' against the *Revue philosophique*.[6] Most philosophers published in both journals, as Paul Vogt's analysis has shown: philosophers seen as close to 'scientistic' approaches published

---

Sciences 38 /4 (2002): 355–69 (359, 358).

3 Anonymous, 'Frédéric Rauh (1848–1909)', *Revue de métaphysique et de morale* 17, supplément de Mars (1909): 2–3 (2).

4 Rauh, however, added that he regretted not being able to 'speak' the language of their 'adversaries', that is the language of one of the special sciences: Perrine Simon-Nahum, 'Xavier Léon/Elie Halévy: Correspondance (1891–1898)', *Revue de métaphysique et de morale* 98 (1993): 3–58 (20, n. 21).

5 Christophe Prochasson, 'Philosopher au XXe siècle: Xavier Léon et l'invention du "système R2M" (1891–1902)', *Revue de métaphysique et de morale* 98 (1993): 109–40 (118).

6 Dominique Merllié, 'Les rapports entre la *Revue de métaphysique* et la *Revue philosophique*: Xavier Léon, Théodule Ribot, Lucien Lévy-Bruhl', *Revue de métaphysique et de morale* 98 (1993): 59–108 (59).

in the *RMM* and 'idealist' philosophers in the *Revue philosophique*.[7] Merllié has pointed out that Lévy-Bruhl's correspondence shows the extent of his collaboration with the *RMM* philosophers. Among other things, he agreed to organize the Paris philosophy conference (in the series of conferences started in 1900 by the *RMM* group), gave papers at the Société française de philosophie and proposed possible speakers.[8] Indeed, not only did Lévy-Bruhl, the editor of the *Revue philosophique* between 1916 and 1939, and the founder of the Institut d'Ethnologie, collaborate with the *RMM* group, but the *RMM* also published Durkheim's articles, lectures and conference papers.[9] Durkheim also gave papers at the Société française de philosophie,[10] and was remembered as a 'very dear collaborator' in the obituary that the journal published.[11] Indeed, as Louis Pinto has pointed out, the Durkheimians' journal, *L'année sociologique*, started as a section of *RMM*, before becoming an independent journal in 1898.[12] Moreover, Lévy-Bruhl's links with Brunschvicg were

7 Paul W. Vogt, 'Identifying Scholarly and Intellectual Communities: A Note on French Philosophy 1900–1939', *History and Theory* 21/2 (1982): 267–78.

8 Merllié, 'Les rapports entre la *Revue de métaphysique* et la *Revue philosophique*', p. 86.

9 Emile Durkheim, 'Représentations individuelles et représentations collectives', *Revue de métaphysique et de morale* 6 (1898): 273–302; Emile Durkheim, 'Pédagogie et sociologie', *Revue de métaphysique et de morale* 11 (1903): 37–54; Emile Durkheim, 'Sociologie religieuse et théorie de la connaissance', *Revue de métaphysique et de morale* 17 (1909): 733–58; Emile Durkheim, 'Les jugements de valeur et les jugements de réalité', *Revue de métaphysique et de morale* 19 (1911): 437–53. The *RMM* also published several of Durkheim's articles posthumously: for a full list, see *RMM*'s special centenary issue, vol. 98 (1983), p. 218.

10 Emile Durkheim et al., 'La détermination du fait moral. Séance du 11 février 1906', *Bulletin de la Société française de Philosophie* 6 (1906): 113–67; Emile Durkheim et al., 'La détermination du fait moral (suite). Séance du 22 mars 1906', *Bulletin de la Société française de Philosophie* 6 (1906): 169–211; Emile Durkheim et al., 'Le problème religieux et la dualité de la nature humaine. Séance du 4 février 1913', *Bulletin de la Société française de Philosophie* 13 (1913): 63. He also attended some of the sessions, see T. Ruyssen et al., 'Pacifisme et patriotisme. Séance du 28 novembre 1907', *Bulletin de la Société française de Philosophie* 8 (1907): 1–30; Gustave Belot et al., 'La morale positive. Examen de quelques difficultés. Séance du 26 mars 1908', *Bulletin de la Société française de Philosophie* 8 (1908): 161–215; Charles Seignobos et al., 'L'inconnu et l'inconscient en histoire. Séance du 28 mai 1908', *Bulletin de la Société française de Philosophie* 8 (1908): 217–47; E. Boutroux et al., 'Science et religion. Séance du 19 novembre 1908', *Bulletin de la Société française de Philosophie* 9 (1908): 19–73; J. Devolvé et al., 'L'efficacité des doctrines morales. Séance du 20 mai 1909', *Bulletin de la Société française de Philosophie* 9 (1909): 193–233; Dominique Parodi et al., 'La notion d'égalité sociale. Séance du 30 décembre 1909', *Bulletin de la Société française de Philosophie* 10 (1909): 53–79; Dr [Jacques-Amédée] Doléris et al., 'L'éducation sexuelle. Séance du 18 février 1911', *Bulletin de la Société française de Philosophie* 11 (1911): 29–51; J. Wilbois et al., 'Une nouvelle position du problème moral. Séance du 2 janvier 1914', *Bulletin de la Société française de Philosophie* 14 (1914): 1–59.

11 Anonymous, 'Emile Durkheim (1858–1917)', *Revue de métaphysique et de morale* 24 (1917): 749–51 (749).

12 Louis Pinto, 'Le détail et la nuance; la sociologie vue par les philosophes dans la *Revue de métaphysique et de morale*, 1893–1899', *Revue de métaphysique et de morale* 98

very strong, including the fact that the former contributed to Brunschvicg's success in obtaining the chair from which he was retiring.[13]

Even a quick glance at the articles and reviews published in these two big philosophy journals would give an idea of the intermingling of historical topics with other philosophical perspectives. In one volume of the *Revue philosophique* (1924) a reader could learn about Pascal from Brunschvicg, read three unpublished letters by Leibniz, keep informed about contemporary Italian thought, and find out about the Fifth International Philosophy Conference that had taken place in Naples (organized by the *RMM* group).[14] If she had an interest in contemporary sciences, there were articles on the crisis of modern physics and on Einstein; Einstein was also discussed in Emile Meyerson's article on 'aprioristic tendency and experience'.[15] An article on the branches of psychology would have introduced our reader to the discipline, while Henri Delacroix would have been an extremely competent guide to the 'psychological conditions of language'.[16] The books reviewed in the 1924 volume, as in the other volumes, were in French, German, Italian and English and covered a great variety of topics, ranging from medieval philosophy to cytology and from rhetoric to the idea of immortality.

If our reader really wanted to keep up-to-date with philosophical research, she would have had to read the *RMM* as well. In the same year, she would have found some articles by the same authors as those in the *Revue philosophique*: Delacroix published in *RMM* another article on language, and so did Meyerson (on relativism) and Brunschvicg, this time on Kant. As it was the bicentenary of Kant's birth, there were other three articles on Kant or Kantianism.[17] The reader interested in sociology

(1993): 141–74 (142). Louis Pinto has calculated the space dedicated to the various disciplines in the first volumes of the *RMM* (1893–99): sociology had 9 per cent of articles, philosophy of science 26 per cent, history of philosophy 14 per cent, psychology 12 per cent, metaphysics 11 per cent and debates on current issues 10 per cent (pp. 141–2).

13 See Lucien Febvre's letter to Henri Berr, December 1926, in Lucien Febvre, *De la Revue de synthèse aux Annales. Lettres à Henri Berr 1911–1954. Etablissement du texte, présentation et notes par Gilles Candar et Jacqueline Pluet-Despatin* (Paris: Fayard, 1997), p. 271.

14 Léon Brunschvicg, 'Pascal savant', *Revue philosophique* 49 (1924): 5–27; G.W. Leibniz, 'Trois lettres', *Revue philosophique* 49 (1924): 5–10; C. Schuwer, 'La pensée italienne contemporaine: L'idéalisme actuel: Giovanni Gentile', *Revue philosophique* 49 (1924): 82–123.

15 Emile Richard-Foy, 'Einstein et sa conception d'un espace fini', *Revue philosophique* 49 (1924): 67–103; Léon Brillouin, 'La crise de la physique moderne', *Revue philosophique* 49 (1924): 11–66; Emile Meyerson, 'La tendance apriorique et l'expérience', *Revue philosophique* 49 (1924): 161–79.

16 H. Delacroix, 'Les conditions psychologiques du langage', *Revue philosophique* 49 (1924): 28–66.

17 H. Delacroix, 'Le fonctionnement psychologique du langage', *Revue de métaphysique et de morale* 31 (1924): 1–17; Emile Meyerson, 'Le relativisme, théorie du réel', *Revue de métaphysique et de morale* 31 (1924): 29–48; Léon Brunschvicg, 'L'idée critique et le système kantien', *Revue de métaphysique et de morale* 31 (1924): 133–203; J. Nabert, 'L'expérience interne chez Kant', *Revue de métaphysique et de morale* 31 (1924): 205–68; L. Robinson, 'Contribution à l'histoire de l'évolution philosophique de Kant', *Revue de métaphysique et de*

could not neglect the *RMM* either: in this very volume, there were articles by Marcel Mauss, sociologically assessing Bolshevism, and Georges Davy, reflecting on industrialization.[18] Those interested in the sciences would have found an article by Federigo Enriques and, in addition to the Meyerson article, a review of his *Explication dans les Sciences* by Parodi.[19]

Historians of philosophy would have had no choice but to keep up with both journals: authors such as Brunschvicg, Delbos, Bréhier, Milhaud, Robin and Wahl published in both, and articles on individual philosophers also appeared in both. Readers interested in specific periods or aspects of the history of philosophy would of course have to read other publications as well, and from 1927, Bréhier's new journal, the *Revue d'histoire de la philosophie*. However, even for historians of philosophy, Bréhier's *Revue* would not have replaced the two great philosophical journals. Indeed, Bréhier only edited it for two years, and its headquarters were moved to Lille in 1933.The most important historians of philosophy published in the two main philosophy journals, and also explained their methodologies and their conceptions of history of philosophy.

The Opposition to the a priori *Systems of Hegel, Comte and Cousin*

The two major philosophical journals would have offered the young historian of philosophy in search of a method extremely important articles in which scholars as such Victor Delbos, Emile Bréhier and Léon Robin set out their views of their discipline. Already at the beginning of the century, the methodology of the history of philosophy had come a long way since Victor Cousin, especially thanks to the reflections and practice of two great philosophers and historians of philosophy, Emile Boutroux (1845–1921) and Charles Renouvier (1815–1903), who, in different ways, also elaborated Kantianism for the use of the history of philosophy. Although their long lives brought them into the twentieth century, and in the case of Boutroux well into the twentieth century, I will not specifically consider them here, although I will refer to their works. Temporal limits, always arbitrary, have to be set, and I assign these two philosophers to the nineteenth century. Boutroux's fundamental work *De la contingence des lois de la nature* appeared in 1874.[20] He did occupy the chair of History of Modern Philosophy at the Sorbonne, but he was appointed in the nineteenth century (1888). He retired in 1907, when Lévy-Bruhl replaced him.

---

morale 31 (1924): 269–353; T. Ruyssen, 'Les origines kantiennes de la Société des Nations', *Revue de métaphysique et de morale* 31 (1924): 355–72.

18 Marcel Mauss, 'Appréciation sociologique du bolchevisme', *Revue de métaphysique et de morale* 31 (1924): 103–32; Georges Davy, 'Le problème de l'industrialisation de l'Etat', *Revue de métaphysique et de morale* 31 (1924): 599–641.

19 F. Enriques, 'La signification et l'importance de l'histoire de la science et l'œuvre de Paul Tannery', *Revue de métaphysique et de morale* 31 (1924): 425–34; Dominique Parodi, 'De l'explication dans les sciences par E. Meyerson', *Revue de métaphysique et de morale* 31 (1924): 585–97.

20 Emile Boutroux, *De la contingence des lois de la nature* (Paris: Presses Universitaires de France, 1990 [1874]).

Victor Delbos was appointed to the Sorbonne chair of Philosophy and History of Philosophy in 1913, but died prematurely only three years later. Three lectures that he dedicated to the conceptions and methods of the history of philosophy were published posthumously in the *RMM*.[21] Delbos's first article provided a systematic criticism of the main models of the history of philosophy, a criticism that was in large part shared by many of his colleagues in the following years, as we shall see below. His first critical target was Hegelianism, which in principle was a major candidate for the method of the history of philosophy, as it regarded philosophy in its historical development, and it closely linked history and philosophy. However, Hegel's philosophy did not satisfy either Delbos or the majority of historians of philosophy until the 1930s. Indeed, even when Hegel's philosophy reached a better currency in France, its model of historical development by and large was not applied to the history of philosophy. Delbos's main criticism was that Hegel did not describe the history of philosophy (as it had really occurred), but he rather chose a series of opinions, often uncharacteristic of a period, and tidily opposed them to one another, in order to satisfy his dialectic views of historical development. Delbos argued that Hegel's history of philosophy was written *a priori*, and as such it was nothing more than the thought of one philosopher projected onto the past.[22] The attentive *RMM* reader would have noticed the similarity between Delbos's criticism and that expressed by Emile Boutroux at the Geneva philosophy conference in 1904, whose proceedings were also published in the *RMM*. There Boutroux had argued that the history of philosophy does not 'conform to a logical plan worked out in advance'.[23]

The Hegelian *a priori* model of development of the history of philosophy remained a target of criticism of important scholars. Bréhier argued that the very idea of a history of all philosophy, from ancient times to the present, was the by-product of the doctrines of the 'progress of the human mind' that emerged at the end of the eighteenth century, and was integrated by Comte and Hegel in their respective philosophies.[24] In their systems, Bréhier objected, the unity of the human mind and the continuity of its development are *a priori* certainties. Against their assumptions, Bréhier argued that the 'immense philological work', which from the mid-nineteenth century had been devoted to the exegesis of known texts and to unearthing new ones, had provided historical evidence against any unitary and progressive view of

21 Victor Delbos, 'Les conceptions de l'histoire de la philosophie', *Revue de métaphysique et de morale* 24 (1917): 135–47; Victor Delbos, 'De la méthode en histoire de la philosophie, I: Les matériaux de la reconstitution historique des doctrines', *Revue de métaphysique et de morale* 24 (1917): 279–89, Victor Delbos, 'De la méthode en histoire de la philosophie, II: Analyse et reconstitution des doctrines', *Revue de métaphysique et de morale* 24 (1917): 369–82.

22 Delbos, 'Les conceptions de l'histoire de la philosophie', pp. 144–5.

23 Emile Boutroux, 'Role de l'histoire de la philosophie dans l'étude de la philosophie', in Edouard Claparède (ed.), *Congrès de Genève 1904* (Geneva: Künding, 1905), pp. 49–68 (57).

24 Emile Bréhier, 'Les postulats de l'histoire de la philosophie', *Revue philosophique* 50 (1925): 48–78 (57). See also Emile Bréhier, 'The Formation of our History of Philosophy', in Raymond Klibansky (ed.), *Philosophy and History. Essays Presented to Ernst Cassirer* (Oxford: Clarendon, 1936), pp. 159–72.

the history of philosophy.[25] Bréhier's insistence on the importance of philological and historical accuracy goes directly against systems that principally rely on a philosophical view of history and on a strong selection of doctrines. He could not accept either nineteenth-century systems, such as Hegel's or Comte's, or eighteenth-century theories such as Condorcet's, although the latter was more acceptable to him because it was open-ended. He thought that all these historiographies presented the evolution of ideas as a necessary progress, and overlooked the importance of the concrete historical and social context.

In his lectures, Delbos discarded not only Hegel's view of history, but also Cousin's, which, although different from Hegel's, was in his view still *a priori*, and employed a handful of general parameters (sensualism, idealism, scepticism and mysticism) not only to characterize all philosophies, but also to describe their succession in time. For Delbos, Cousin's evaluation of doctrines was 'vague and arbitrary', and founded on the misconception according to which every philosophical system originates in a 'kind of general element'. For Delbos it made little sense to apply the label 'sensualist' to such diverse and historically distant philosophies as Epicurus', Locke's and Condillac's; the category of 'idealism' was even less useful as it included doctrines which were not only different but indeed incompatible. Once again, an exclusively philosophical and *a priori* approach to the history of philosophy for him would not give an acceptable representation either of the single doctrines or of their succession in time. His criticism of anti-historical approaches did not spare Charles Renouvier, who, while opposing Hegelianism and eclecticism, nevertheless in his view still lacked historical accuracy. His presentation of the history of philosophy as a collection of dilemmas, Delbos argued, might be philosophically interesting, but it has no historical truth.[26]

Similarly to Delbos, Bréhier dismissed Cousin's methodology, writing that it was like that of the botanist who classifies plants;[27] elsewhere he remarked that the 'oratorical spiritualism', which was the legacy of Cousin's school, aimed to entertain rather than to demonstrate anything.[28] However, for Bréhier, who wrote much of his work after Delbos's death, eclecticism was no longer a worthy target; his critical remarks went rather to Hegelian and Comtian models, which enjoyed more enduring legacies.

Bréhier suggested that Delbos did not write a general history of philosophy probably because of his aspiration to precision and historical accuracy.[29] Despite the fact that Bréhier did write a monumental history of philosophy,[30] he was very sensitive

25 Bréhier, 'Les postulats de l'histoire de la philosophie', p. 72.

26 Delbos, 'Les conceptions de l'histoire de la philosophie', pp. 146–7.

27 Bréhier, 'Les postulats de l'histoire de la philosophie', p. 65.

28 Emile Bréhier, *La philosophie et son passé* (Paris: Presses universitaires de France, 1950 [1940]), p. 11. Cresson was not more flattering: he concluded his exposition of Cousin's philosophy by judging it a 'surprisingly simple doctrine … too simple, no doubt, and a little naïve': Cresson, *Les courants de la pensée philosophique française*, vol. 2, p. 98.

29 Bréhier, 'Les postulats de l'histoire de la philosophie', pp. 73–4.

30 Emile Bréhier, *Histoire de la philosophie, vol. I: L'antiquité et le Moyen Age. I: Introduction, Période hellénique* (Paris: Alcan, 1926); Emile Bréhier, *Histoire de la philosophie, vol. I: L'antiquité et le Moyen Age. II: Période hellénistique et romaine* (Paris:

to the difficulties of such an enterprise, for the reasons that Delbos mentioned or implied. Bréhier did not hesitate to generalize about the philosophy of his time, which he regarded as having a clear aversion to Hegelian or positivistic 'grand constructions'.[31] Among the authors who limited themselves to a period or a specific problem, he cited not only French ones, but also Germans, therefore not siding on this occasion with Bergson, who had claimed that the opposition to systems, like the Hegelian or Kantian one, was a specific trait of French philosophy.[32] Bréhier would have agreed with Brunschvicg, for whom 'the France of Auguste Comte has no ground either for envying or for condemning anything in either the Germany of Hegel or the England of Herbert Spencer'.[33]

Bréhier's and Delbos's vindication of historical accuracy was shared by Etienne Gilson, who had been a student of both Delbos and Lévy-Bruhl. Indeed, by his own admission, Gilson learned from Lévy-Bruhl's books on Comte and Jacobi to present philosophers' thought from their own points of view, that is in a historically accurate manner.[34] His own method of careful interpretation of texts, which Jorge Gracia has called 'enlightening gloss',[35] not to mention his own Thomistic philosophy, were therefore at odds with any of these systems that had shaped the history of philosophy in the nineteenth century. Brunschvicg, as we shall see, did present the history of philosophy as progress, but for him its progress is empirically observed, rather than logically worked out.

The importance of historical accuracy appeared to be stressed by many important scholars. But how should this be exercised, and more importantly, what role should it play? By reading Bréhier, one would think that the historical and social circumstances in which philosophical works appeared were to play an important role in the interpretation of the texts themselves, as he emphasized in the presentation of his new journal, *Revue d'histoire de la philosophie*.[36] He stressed that historians

---

Alcan, 1927); Emile Bréhier, *Histoire de la philosophie, vol. I: L'antiquité et le Moyen Age. III: Moyen Age et Renaissance* (1928); Emile Bréhier, *Histoire de la philosophie, vol. II: La philosophie moderne. I: Le dix-septième siècle* (Paris: Alcan, 1929); Emile Bréhier, *Histoire de la philosophie, vol. II: La philosophie moderne. II: Le dix-huitième siècle* (Paris: Alcan, 1930), Emile Bréhier; *Histoire de la philosophie, vol. II: La philosophie moderne. III: le XIXe siècle. Période des systèmes (1800-1950)* (Paris: Alcan, 1932); Emile Bréhier, *Histoire de la philosophie, vol. II: La philosophie moderne. IV: Le XIXe siècle après 1950. Le XXe siècle* (Paris: Alcan, 1932).

31 Bréhier, 'Les postulats de l'histoire de la philosophie', p. 70.

32 Henri Bergson, 'La philosophie française (Tableau récapitulatif destiné à l'Exposition de San Francisco)', *Revue de Paris* 22/3 (1915): 236–56 (254).

33 Léon Brunschvicg, 'History and Philosophy', in Klibansky (ed.), *Philosophy and History*, pp. 30–31.

34 'Une heure avec Etienne Gilson' [1925], in Frédéric Lefèvre, *Une heure avec ...* (Paris: Gallimard, 1924–27), p. 65.

35 Jorge J. E. Gracia, 'The Enlightening Gloss: Gilson and the History of Philosophy', in Peter A. Redpath (ed,), *A Thomistic Tapestry: Essay in Memory of Etienne Gilson* (Amsterdam and New York: Rodopi B.V., 2003), pp. 1–11.

36 Anonymous [E. Bréhier], 'Notre programme', *Revue d'histoire de la philosophie* 1/1 (1927): 1–4.

of philosophy cannot neglect the history of other disciplines, and, when they read a text, they should take into account its contemporary culture and events.[37] However, the same reader could have been puzzled by Bréhier's almost exclusive focus on philosophical texts in his own work.

On the other hand, from the pages of the *Revue philosophique*, Robin, while allowing for a certain usefulness of 'problems of chronology, or authenticity, or sources', judged that the essence of the work of the historian of philosophy was that of 'listening to the voice' of the author who is persuaded to have captured a philosophical truth. In other words, Robin advocated a direct dialogue with past authors, indeed the historian of philosophy for him should re-live the thoughts of past philosophers.[38] Once again, the reader could have been a little confused by Robin's own attention to philology in his own approach to texts.

These few articles would have already raised some questions: does the history of philosophy have an independent history, or is its history linked to other disciplines and to political and social history? The answer to these questions implies a different view of philosophy as an activity, but also of the historians of philosophy's craft: should they limit themselves to philosophical texts or expand their research? Is it more important to establish links between philosophical texts across time, or rather to connect a text with contemporary texts and events? At a theoretical level, different answers to these questions would imply different conceptions of 'history'. One is history as 'vertical' history, as Bréhier put it, or as chronological development, which may or may not take the form of progress. This history has a direction and it leads to the present. However, if the aim of the historian of philosophy is the direct dialogue with a past text, 'history' seems to be something rather different. Here the chronological distance is not very important, as it can be bridged by philosophical activity: it can be taken into account in order better to interpret the text, but in the end time is eliminated by the interpreters' direct dialogue with their texts.

Another issue underlying these theoretical discussions of the history of philosophy is that of its goal. What do historians of philosophy intend to achieve in their activity? Is it the correct interpretation of past doctrines? Or is it rather to philosophize by reflecting on philosophy's 'past efforts', as Robin put it? Bréhier also suggested that the history of philosophy is a reflection that reason makes on itself, and that it is a way of 'knowing oneself'.

The aims, nature and methods of the history of philosophy were not just proposed in journal articles and introductions to the history of philosophy, but were also the topics of face-to-face discussions. The Société française de philosophie was without doubt where the most important discussions took place. The examination of a few of them will start answering some of the questions about history of philosophy.

37 Ibid.; Bréhier, 'L'histoire de la philosophie dans l'enseignement', pp. 211–2.

38 Léon Robin, 'L'histoire et la légende de la philosophie', *Revue philosophique* 60 (1935): 161–75 (174).

Debates about the Notion of the History of Philosophy

The Société française de philosophie: A Place for Discussion

The Société française de philosophie was an extremely important place of debate for philosophers (in the broadest sense) in Paris in the first half of the twentieth century. Founded in 1901 by Xavier Léon and André Lalande,[39] it organized talks by prominent scholars, and published the proceedings in its *Bulletin*. As Xavier Léon narrates in his letters to Elie Halévy,[40] co-editor, with himself, of the *Revue de métaphysique et morale*, the birth of the Société was decided on the morning of 29 November 1900, by Xavier Léon, Henri Bergson, Gustave Belot, André Lalande, Victor Delbos, Louis Couturat and Léon Brunschvicg.[41] Between the beginning of the twentieth century and the First World War, the *Bulletin* recorded seven or eight meetings per academic year with few exceptions, with roughly monthly regularity between October or November and July, with no meetings in the Easter break, and sometimes none in June. The war made it impossible to continue regular publications, which however resumed after the war, with around five meetings a year.

As an off-shoot of the *Revue de métaphysique et de morale*, the Société over the years discussed a great variety of philosophical issues and hosted not only philosophers, but also social scientists, psychologists and scientists, including Albert Einstein and Louis de Broglie.[42] Among the sociologists, as mentioned above, Durkheim himself

39 See the Société française de philosophie webpage http://www.sofrphilo.fr/ (accessed 1 June 2007).

40 Elie Halévy (1870–1937), son of a writer, admitted to the ENS in 1899, *agrégé*, wrote a doctorate on Bentham and philosophical radicalism in England (1901); he lectured at the Ecole des sciences politiques, but refused a post at the Sorbonne. His father was Jewish, and his mother Protestant. See Nicolas Roussellier 'Halévy (Elie)' in Julliard and Winock (eds), *Dictionnaire des intellectuels français,* pp 582-3. Fabiani, *Les philosophes de la République*, p. 94; Simon-Nahum, 'Xavier Léon/Elie Halévy: Correspondance (1891–1898)', p. 6.

41 Xavier Léon, Letter to Halévy, 30 November 1900, quoted in Henri Bergson, *Lettere a Xavier Léon e ad altri* (Naples: Bibliopolis, 1992), pp. 66–8, n. 1. The mathematician and Leibniz scholar Louis Couturat (1868–1914) was another connection from the *lycée* Condorcet, where Couturat had been, like Brunschvicg, Léon and Halévy, a pupil of the philosopher Alphonse Darlu. Couturat was admitted to the ENS in 1888, was a philosophy *agrégé*, but proceeded to gain a degree in mathematics. He lectured at the University of Toulouse and Caen, and was nominated as supply professor at the Collège de France in 1905, to replace Bergson. See Simon-Nahum, 'Xavier Léon/Elie Halévy: Correspondance (1891–1898)', p. 14, n. 1. As Prochasson has stressed, Couturat created a link between scientific milieus and the *RMM* group, although Halévy wrote to Léon that he found Couturat 'insufferable', see Prochasson, 'Philosopher au XXe siècle', p. 112. Gustave Belot, *normalien*, in 1899 replaced Lévy-Bruhl at the *lycée* Louis-le-Grand, became Inspecteur de l'Académie de Paris in 1911 and Inspecteur général de l'instruction publique in 1913 (Archives nationales AJ/16/952).

42 Louis de Broglie et al., 'Le déterminisme et la causalité dans la physique contemporaine. Séance du 12 novembre 1929', *Bulletin de la Société française de Philosophie* 29/5 (1929): 141–60, Albert Einstein, et al., 'La théorie de la relativité. Séance du 6 avril 1922', *Bulletin de la Société française de Philosophie* 22 /3 (1922): 91–113. Einstein attended de Broglie's session.

presented papers and attended sessions; and so did Célestin Bouglé.[43] Lévy-Bruhl presented both his *La mentalité primitive* and *L'âme primitive* to the members of the Société,[44] and Marcel Mauss contributed to discussions.[45] Psychologists were also invited, including Henri Piéron, editor of *L'Année psychologique* and director of the laboratory of psychology at the Sorbonne,[46] and Jean Piaget.[47] The Société also promoted and supported the production of the now classic *Vocabulaire technique et critique de la philosophie.*[48] André Lalande and his collaborators (including Victor Delbos, Gustave Belot and Louis Couturat) would produce sets of entries that were first discussed by the core group, and then sent to the Société membership, including the foreign correspondents. The Société would then hold a meeting to discuss the new entries;[49] the record of both the written corrections and the discussions published in the *Bulletin* testifies that the *Vocabulaire* was a truly collective enterprise.

The record of the direct interactions of the main protagonists of these debates is an invaluable document which enables the historian of this period to form a more complete picture of philosophy than the individual articles and books would. It is obviously impossible to discuss all the sessions relevant to the topic of the present book, but there were a few that are cornerstones in the debates about the historicity of the mind and the history of philosophy. The papers that the historians of philosophy Lucien Lévy-Bruhl, Léon Brunschvicg and Léon Robin read at sessions that took place in the 1920s and 1930s are particularly relevant. I shall examine Lévy-Bruhl's and Brunschvicg's views in detail in the next chapter. First, I shall examine the set of views of the history of philosophy that I call 'mainstream', by focusing on the paper on the notion of the history of philosophy that Robin delivered at a meeting of the Société, and the discussion that followed it.

---

43 Célestin Bouglé et al., 'La sociologie du Proudhon', *Bulletin de la Société française de Philosophie* 12 (1912): 169.

44 Lucien Lévy-Bruhl et al., 'La mentalité primitive. Séance du 15 février 1923', *Bulletin de la Société française de Philosophie* 23 (1923): 17–48, Lucien Lévy-Bruhl et al., 'L'âme primitive. Séance du 1er juin 1929', *Bulletin de la Société française de Philosophie* 29 (1929): 105–37.

45 See Lévy-Bruhl et al., 'L'âme primitive'; J. Vendryès et al. 'Le progrès du langage. Séance du 30 novembre 1922', *Bulletin de la Société française de Philosophie* 22 (1922): 151–69; Paul Mus et al., 'La mythologie primitive et la pensée de l'inde', *Bulletin de la Société française de Philosophie* 37 (1937): 83–126; Louis Weber et al., 'Liberté et langage. Séance du 23 juin 1921', *Bulletin de la Société française de Philosophie* 21 (1921): 75–106.

46 Charle and Telkès, *Les professeurs du Collège de France,* p. 204; H. Piéron et al., 'La notion d'instinct. Séance du 28 mai 1914', *Bulletin de la Société française de Philosophie* 14 (1914): 301–36; H. Piéron et al., 'L'étude biologique de la mémoire. Séance du 12 janvier 1911', *Bulletin de la Société française de Philosophie* 11 (1911): 1–27.

47 J. Piaget et al., 'Les trois systèmes de la pensée de l'enfant. Séance du 17 mai 1928', *Bulletin de la Société française de Philosophie* 28 (1928): 97–141.

48 André Lalande, *Vocabulaire technique et critique de la philosophie* (2 vols; Paris: Presses Universitaires de France, 1999 [1926]).

49 See Lalande's report in Louis Couturat, 'IIme Congrès de Philosophie, Genève, comptes rendus critiques, II: Logique et philosophie des sciences', *Revue de métaphysique et de morale* 12 (1904): 1037–1077 (1071–2).

The Notion of the History of Philosophy: Mainstream Views of the History of Philosophy

The session of the Société française de philosophie of 25 April 1936 was introduced by a talk by Léon Robin on the 'notion of history of philosophy'.[50] The Sorbonne Professor of History of Ancient Philosophy aimed to develop the view that he had already presented on the pages of the *Revue philosophique* a year earlier.[51] At the discussion at the Société française de philosophie, he chose to concentrate in particular on his defence of the philosophical status of the history of philosophy. He did concede that the first stage of the historian of philosophy's work involved 'ordinary' historical methods, including philology.[52] This preliminary work, he explained, clarifies the doctrine under study and its reception. However, in his view the authentically philosophical task of the historian of philosophy was beyond the simple historical and philological analysis; it was rather a work of 'constructive synthesis' and 'assimilation' of the doctrine.[53] For Robin, the philosophical reading of doctrines that he saw as the essence of the history of philosophy is based on a direct and creative relationship between the reader and the text. He insisted that there is nothing 'dead' in the history of philosophy, for a reader can 'resurrect' any text. Robin appeared here to defend an anachronistic approach to past texts, as for him the time and context in which a text is produced is not necessarily relevant to it. And indeed he claimed that the history of philosophy is 'utopian' and 'uchronic'. Robin's use of the latter term is very interesting. At face value, it tells us that he thought the historian of philosophy might draw the lines of historical development rather than being constrained by them. His use of this term, however, was also an appeal to the authority of Charles Renouvier. The latter coined it and, in 1876, published a volume entitled *Uchronie (L'utopie dans l'histoire): Esquisse historique apocryphe du développement de la civilisation européenne tel qu'il n'a pas été, tel qu'il aurait pu être*.[54] The idea behind Renouvier's book is the rejection of objective laws of historical development. Consistent with Renouvier's broader doctrine, Robin rejected the idea of the history of philosophy as a development governed by laws or exhibiting progress. He rather saw doctrines as having an intrinsic and individual value, which is not linked to time and cultural context.

Robin also followed Renouvier in vindicating the philosopher's freedom to believe in one doctrine rather than another. Renouvier argued that philosophical ideas

---

50 Léon Robin et al., 'Sur la notion d'histoire de la philosophie. Séance du 25 avril 1936', *Bulletin de la Société française de Philosophie* 36 (1936): 101–40. Among those who took part in the discussion were Léon Brunschvicg, Isidore Lévy, Paul Etard, Jean Baruzi, Jean Wahl, Pierre Ducassé, Alexandre Koyré and Henri Berr.

51 Robin, 'L'histoire et la légende de la philosophie'.

52 Robin in ibid., p. 103.

53 Robin in ibid., p. 107.

54 Charles Renouvier, *Uchronie (L'utopie dans l'histoire): Esquisse historique apocryphe du développement de la civilisation européenne tel qu'il n'a pas été, tel qu'il aurait pu être* (Paris: Bureau de la critique philosophique, 1876).

are based on a free choice,[55] rather than on evidence. If this were not the case, he argued, then an agreement on the great philosophical questions (which he believed to be the same throughout the history of philosophy) would have been reached, with a consequent loss of freedom of choice. He thought that belief rather than evidence guides the elaboration of, and provides support for, philosophical ideas, and that belief is grounded on a moral choice. The above-mentioned article, which Robin published one year before this talk, is even more reminiscent of Renouvier's positions on the history of philosophy. There Robin claimed that 'true' and 'false', when applied to philosophical doctrines, express a personal preference, and 'an original and free act of thought'. For him, historians of philosophy do not mainly engage with past doctrines in order to assess their truthfulness, but rather in order to develop their own beliefs and philosophy. Robin thought that 'disinterestedness' was not the historian of philosophy's business.[56]

Given his emphasis on belief rather than proof, and his defence of the ultimately ahistorical character of doctrines, it comes as no surprise that Robin viewed the history of religion as the closest to the history of philosophy, and the history of science as the furthest. In Robin's argument, the difference between the history of philosophy and the history of science depends on the role that truth plays in these two disciplines. Robin assumed that in science truth is absolutely separated from error; as a consequence, doctrines are expelled from the history of science at the point in which they are deemed to be false. They cannot be employed again. By contrast, for Robin philosophical doctrines can be given new life and relevance by later philosophers. The difference that Robin saw between the attitudes towards the past that historians of science and historians of philosophy respectively held had important consequences for the overall structure of these two histories. From his claims, it follows that the history of science exhibits a progress, whereas the history of philosophy does not.

The discussion that followed Robin's talk highlighted the two ideally opposite, but in fact often linked, approaches to the history of thought: one aimed at historically and philological accuracy, and the other aimed at the creative interpretation of past philosophies. Scholars who defended either approach were present in the audience. Interestingly, the advocates of the philological approach came from the ranks of the historians of religion, whom Robin had seen as the closest to his idea of an ultimately anachronistic view of past doctrines. Indeed, Isidore Lévy, Professor of Ancient History and the Semitic Orient at the Collège de France, was keen to differentiate his own historical approach from any empathic relationship with past texts.[57] He

55 Renouvier developed this thesis in Charles Renouvier, *Esquisse d'une classification systématique des doctrines philosophiques* (2 vols; Paris: Fischbacher, 1885), vol. 2, 'Sixième partie – sixième opposition: L'évidence; la croyance', pp. 1–126; see also Gueroult, *Dianoématique, livre I: L'histoire de l'histoire de la philosophie, 3: en France de Condorcet à nos jours* (Paris: Aubier, 1988), ch. 33. For Renouvier's view of science, see Warren Schmaus, 'Renouvier and the Method of Hypothesis', *Studies in History and Philosophy of Science* 38 (2006): 132–48.

56 Robin, 'L'histoire et la légende de la philosophie', p. 162.

57 Isidore Lévy (1871–1959), after teaching at the Ecole pratique des hautes études (1905–38), where he became Director of Studies of History of the Ancient East, was elected

declared that the approach that Robin denigrated as 'palaeontological' was the very one he adopted. He regarded his work as that of reconstructing the past from the evidence available to us. Robin's insistence on the philosophical significance of Lévy's book on Pythagoras,[58] if anything, irritated his interlocutor. The Professor of History of Religions at the Collège de France, Jean Baruzi,[59] was more conciliatory, and expressed his approval of the similarity between the history of philosophy and the history of religions, but his focus was still on historical reconstructions and accuracy. He regarded any special history as a small part of history as a whole. In this, he was obviously supported by Henri Berr, the *ideologue* of historical synthesis.[60]

Paul Etard[61] presented his view of history as intellectual development, which was the very opposite of Robin's, but would have pleased positivists and social scientists. He drew a clear distinction between two attitudes towards past philosophies. He explained that one attitude was to determine the stages of the development of the human mind. In order to write a history of the mind, there is no reason for a scholar to consider only philosophical texts; indeed religious doctrines could be just as important. The other attitude, which Etard attributed to Robin, and in general to philosophers, including Boutroux and Lachelier, consisted in finding one's thought in past philosophies, and to seek interesting themes and ideas in past texts in order to stimulate one's own thought. Where Etard's sympathies lay was obvious: he found many philosophical interpretations 'shocking' for the historian, and concluded by saying that all definitions of history are opposed to such attitude.[62] More 'pure' historians, however, did not emphasize the 'stages' of the human mind, but rather the study of historical periods in all their aspects, so connecting philosophy to all other disciplines, and intellectual history with all other special histories. Baruzi defended this view, and so did Henri Berr in the same session.

Robin's defence of an empathic relationship with past texts found convinced supporters among philosophers. One such supporter, present in the audience, was Jean Wahl.[63] Wahl, who after the war would become the secretary of the Société and the editor of the *Revue de métaphysique et de morale*, had already published important works on the history of philosophy, including *Le malheur de la conscience*

---

to a chair at the Collège de France (1933–45). His parents were Jewish trades people, and his wife's sister married Marcel Mauss's brother: Charle and Telkès, *Les professeurs du Collège de France*, pp. 139–40.

58 Isidore Lévy, *Recherches sur les sources de la légende de Pythagore* (Paris: Leroux, 1926).

59 Jean Baruzi (1881–1953), from a well-off family of Italian origin, was elected to a Collège de France chair the same year as Lévy (1933); see Charle and Telkès, *Les professeurs du Collège de France*, pp. 29–30; *Annales de l'Université de Paris* 2 (1927): 295.

60 Robin et al., 'Sur la notion d'histoire de la philosophie', p. 133.

61 Paul Etard was ENS librarian (he had replaced Lucien Herr), and had previously taught at the *lycée* in Strasbourg; see Ecole normale supérieure, 'Rapport du directeur', *Annales de l'Université de Paris* 2 (1927), p. 213.

62 Etard in Robin et al., 'Sur la notion d'histoire de la philosophie', p. 123.

63 Wahl in ibid., p. 123–4.

dans la philosophie de Hegel.[64] This latter work, which played a major role in reviving interest in Hegel among French philosophers, should not lead us to believe that Wahl's Hegelianism translated into a view of the history of philosophy similar to Hegel's, or that he advocated any system or laws of historical development. Indeed, his work provided a personal reading of Hegel, focused on the 'unhappy consciousness'; it was inspired by Hegel's early writings, filtered through existentialist philosophy and sensitivity. Wahl aimed at disclosing the 'tragic, romantic, religious element' of Hegel's philosophy.[65] His method in reading Hegel was the same as that he defended in this session of the Société: a direct dialogue with the text, in which empathy plays a major role, and which is aimed at truth in the same way as poetry is.

Scholars such as Robin and Wahl regarded past philosophers ultimately as their contemporaries, and believed that they expressed ideas that the modern reader could choose to embrace, regardless of time and place. Their approach was prevalent among historians of philosophy, who conceived of the history of philosophy as a collection of great texts and ideas, through which modern scholars developed their own philosophical ideas and doctrines.

At first sight, Robin's position seems rather different from Bréhier's, for the latter insisted on historical and philological accuracy, while Robin appeared to downplay its importance. In fact, although their philosophical views, for instance about truth, were indeed rather different, they both valued historical accuracy and vindicated an empathic relationship with texts. What made their views on these issues appear discordant was their different polemical targets and objectives. Robin aimed to defend the philosophical status of the history of philosophy, whereas Bréhier, invoking historical accuracy, intended to restore to philosophy the real variety of ideas that Hegel's, Comte's and indeed Cousin's approaches had erased.[66]

---

64 Jean Wahl, *Les philosophies pluralistes d'Angleterre et d'Amérique* (Paris: Alcan, 1920); Jean Wahl, *Etude sur le Parménide de Platon* (Paris: Rieder, 1926); Jean Wahl, *Vers le concret: Etudes d'histoire de la philosophie contemporaine* (Paris: Vrin, 1932); Jean Wahl, *Le malheur de la conscience dans la philosophie de Hegel* (Paris: Rieder, 1929).

65 Wahl, *Le malheur de la conscience dans la philosophie de Hegel*, p. 9.

66 Similarly, Delbos appealed to historical contingencies to dispel the impression that his narration of the route that philosophy took from Kant to the post-Kantians could be interpreted as a development of ideas moved by an internal logic, and substantially independent of both their authors and the milieu in which they appeared. He warned his students at the very beginning of his lectures on this topic that his account must not be read as a working out of an internal dialectic, precisely because 'psychological and social' causes made some developments possible and barred others: see Victor Delbos, 'Les facteurs kantiens de la philosophie allemande du commencement du XIXe siècle', *Revue de métaphysique et de morale* 26 (1919): 569–93 (570). The other lectures were published posthumously: Victor Delbos, 'Les facteurs kantiens de la philosophie allemande de la fin du XVIIIe siècle et du commencement du XIXe', *Revue de métaphysique et de morale* 27 (1920): 1–25; Victor Delbos, 'Les facteurs kantiens de la philosophie allemande de la fin du XVIIIe siècle et du commencement du XIXe', *Revue de métaphysique et de morale* 28 (1921): 27–47; Victor Delbos, 'Les facteurs kantiens de la philosophie allemande de la fin du XVIIIe siècle et du commencement du XIXe', *Revue de métaphysique et de morale* 29 (1922): 157–76. They were also collected in one volume: Victor Delbos, *Des Kant aux postkantiens* (Paris: Aubier, 1992 [1940]).

However, Bréhier was not only interested in historical accuracy, nor did he believe that philosophical texts only have historical significance. In fact, he argued that the philosopher does not aim to uncover why a doctrine emerged, nor its place in the historical development, but rather whether a doctrine is true.[67] He opposed historical relativism;[68] for him, the truth of a doctrine does not depend on the historical and social circumstances of its emergence, as a consequence it can be grasped by readers regardless of their distance in time. He believed that a direct relation between reader and philosophical text was not only possible, but indeed was the very essence of the historian of philosophy's work. On this point, he wrote that he had learned from Bergson to consider as contemporaries authors who lived even centuries earlier than he did.[69] This relationship between modern philosopher and past texts was for Bréhier ultimately subjective; he represented the history of philosophy as a series of 'appeals' that make the reader's consciousness 'vibrate' with sympathy.[70] Despite their differences, Bréhier, like Robin, defended a direct relationship with texts and the ultimate independence of philosophy from historical, social and economic circumstances.[71]

The Société française de philosophie also hosted talks that appeared to challenge both the view of philosophical ideas as potentially timeless, and their special status in comparison for instance to scientific ideas. Indeed, one of the Société's founders, Léon Brunschvicg, practised the history of philosophy in a manner that seemed to contradict those views, when he embarked on a study of the mind through the history of philosophy and indeed the history of science; he appeared to erase the difference between these two histories, which was seen as fundamental by many historians of philosophy. Lévy-Bruhl also embarked on a project of the history of the mind through the examination of texts, and he too appeared to undermine the special and timeless status of philosophical texts. I shall examine their positions in the following chapter.

Before that, there is still something to be investigated about the methodology of the 'mainstream' historians of philosophy. In the next section, I shall discuss one of the most important sources of inspiration for those historians of philosophy who regarded philosophy as something distinct from any other cultural phenomenon, and its history as different from the history of other disciplines, in its content and methods. The philosopher who probably enjoyed the most widespread fame in the early twentieth century, and to whom historians of philosophy from Bréhier to Gilson acknowledged their debt, Henri Bergson, presented his view of the history of philosophy at one of the international conferences that the *RMM* group promoted.[72]

---

67 Emile Bréhier, *Transformation de la philosophie française* (Paris: Flammarion, 1950), p. 7.

68 Bréhier, *La philosophie et son passé*, p. 43.

69 Bréhier, Letter to Gueroult [1948], quoted in Gueroult, *Dianoématique, livre I: L'histoire de l'histoire de la philosophie, 3*, p. 939.

70 Bréhier, *La philosophie et son passé*, p. 44.

71 Bréhier described his view of history of philosophy as rationalist and individualist: Emile Bréhier, 'Comment je comprends l'histoire de la philosophie', *Les Etudes philosophiques*, nouvelle série, 2 (1947): 105–14 (112).

72 See 'Une heure avec Etienne Gilson' [1925] in Lefèvre, *Une heure avec ...* , p. 65.

Bergson at the International Congress of Philosophy and the Theoretical Foundations of Mainstream Conceptions of the History of Philosophy

The meetings of the Société française de philosophie were held in Paris, and, despite the numerous foreign scholars invited, it was substantially a French affair. The *RMM* group, however, did not overlook the international dimension of their enterprise, and, already at the very end of the nineteenth century, started planning international congresses of philosophy. The organization of the first one was entrusted to Louis Couturat and took place in Paris in 1900.[73] The second was held in 1904 in Geneva, the third in 1908 in Heidelberg, the fourth in 1911 in Bologna, and then in Naples (1924) and Harvard (1926), followed by those of Oxford, Prague and Paris again in 1937. After the war, it travelled to many cities, including Amsterdam, Brussels, Florence, Mexico City and Vienna.[74] The history of philosophy was well represented in these international conferences from the beginning: at the 1900 congress, such scholars as Emile Boutroux, Henri Delacroix, Victor Delbos, Gustave Belot, Paul Tannery, René Bertholot and Victor Brochard were among those who chose to deliver their papers in the sessions dedicated to the history of philosophy, and so did Wilhelm Windelband, Jean Benrubi, Xavier Léon and Léon Brunschvicg, among many others, in 1904. Xavier Léon also delivered his papers in these sessions in Heidelberg (1908) and Bologna (1911).[75]

The papers of interest to historians of philosophy were certainly not limited to the sessions explicitly dedicated to this branch of philosophy. The historians of philosophy who attended the 1911 conference in Bologna should not have missed Henri Bergson's plenary session, especially if they were interested in the methodology underlying their research. It would, however, have been difficult to guess the relevance of Bergson's paper for this branch of philosophy from its title, which simply read: 'Philosophical intuition'.[76] Moreover, although one of his

73 Prochasson, 'Philosopher au XXe siècle: Xavier Léon et l'invention du "système R2M" (1891–1902)'. For a discussion of the Congresses of Philosophy that took place between 1900 and 1911, see Gaspare Polizzi, *Tra Bachelard e Serres: aspetti dell'epistemologia francese del Novecento* (Messina: Armando Siciliano, 2003), ch. 1.

74 For a list of the talks delivered in these conferences, see Lutz Geldsetzer, *Bibliography of the International Congresses of Philosophy/Bibliographie der Internationales philosophie Kongresse: Proceedings/Beiträge 1900–1978* (Munich, New York, London and Paris: Saur, 1981).

75 Brunschvicg presented his talks both in the history of philosophy sessions and in others, such as the 'General philosophy and metaphysics' one (1900, 1908), as well as being one of the plenary speakers.

76 Henri Bergson, 'L'intuition philosophique', in International Congress of Philosophy (ed.), *Atti del IV Congresso Internazionale di filosofia, Bologna MCMXI, vol. I* (Genova: Formiggini, 1912 [1911]), pp. 174–92; also published in *RMM*: Henri Bergson, 'L'intuition philosophique', *Revue de métaphysique et de morale* 19 (1911): 809–27. It is impossible to know who was physically in the audience, apart from the chairs (Oswald Külpe and Alfred William Benn) and the other two speakers (Svante Arrehenius and Paul Langevin). However, among the participants of the Bologna conference (515 in total, of whom 177 gave papers) were Boutroux, Durkheim, Poincaré, Rey, Parodi, Lalande, Masson-Oursel, Léon, Lévy-

courses at the Collège de France was on the history of philosophy,[77] Bergson was not considered a historian of philosophy. In any case, not only did most historians of philosophy consider themselves to be first of all philosophers, but they also might not have wanted to miss his paper, for his importance in the philosophical world of the time cannot be overstated, and his influence extended to all branches of philosophy.

The conception of the history of philosophy that Bergson put forward in Bologna differed from that which he had proposed in his *Evolution créatrice* only a few years earlier. In the latter work, Bergson had presented his own philosophy as a new departure, indeed as authentic philosophy, in contrast with past philosophical systems, which had not really emancipated themselves from science.[78] By contrasting his own true philosophy with past philosophies, he devalued the latter. In Bologna, he presented a much more positive view of the history of philosophy. What had not changed, however, was his conception of the relation between philosophy and science. Bergson regarded philosophical knowledge as distinct from science and common sense both in its object and in its method. For him, science is applied to 'inert matter' and, by using the intellect, can only perceive what is immobile, whereas philosophy, by using intuition, is able to grasp life and movement. What science and common sense call movement and change for Bergson is only a succession of snapshots, of immobile and lifeless moments. He called this the 'cinematographic' illusion: we think we see movement; instead, we only see stills in rapid succession. Philosophical intuition is for him capable of going beyond this type of knowledge and grasping the fluid movement of time (*durée*).[79]

For Bergson philosophy springs out of an original, personal and ultimately ineffable intuition. In his view, philosophers do need concepts and images in order to give shape to their intuitions and to communicate them: they have no choice but to use the language and concepts of their times and address the same issues as others do. However, their images and concepts are not for him the essence of their thoughts, but only their external form. From this conception of the essence of philosophy as original intuition, Bergson drew several consequences, which he presented to the participants of the Bologna conference. First of all, the ordinary work of placing a philosophical text historically, and of tracing sources and influences, may be useful, or even necessary in an instrumental way, but will never tell us what a certain philosophy is about, or what is original about it. The work of the historian of philosophy is to arrive as close as possible to the 'infinitely simple' element that is

---

Bruhl, Delbos, Couturat, Louis Weber, Jean Benrubi, Blondel, Brunschvicg, Benedetto Croce, Guido De Ruggiero and Wilhelm Windelband. The organizer of the conference was Federigo Enriques: see International Congress of Philosophy, *Atti del IV Congresso Internazionale di filosofia, Bologna MCMXI* (3 vols; Genova: Formiggini, 1912–16), vol. 1, pp. 338–53, pp. 355–56; vol. 3, pp. 679–82.

77 Bergson, 'L'intuition philosophique', p. 178.

78 For the contrast between Bergson's two conceptions of history of philosophy, see Gueroult, *Dianoématique, livre I: L'histoire de l'histoire de la philosophie, 3*, ch. 34: 'Bergson'.

79 Henri Bergson, *L'évolution créatrice* (Paris: Alcan, 1907). For a clear discussion of Bergson's conception of intuition, see Gilles Deleuze, *Bergsonism* (trans. Hugh Tomlison and Barbara Habberjam; New York: Zone Books, 1991 [1966]), ch. 1: 'Intuition as method'.

at the core of the philosophy under study, and that the author has never managed to express directly. If the original author could not express it directly, historians cannot certainly do it either, but, Bergson explained, they can still provide an 'intermediary image', which lies somewhere in between the simplicity of the original intuition and the complexity of the form in which the author has chosen to present it.[80]

Bergson provided some examples of the method that he recommended: an interpreter of Spinoza's *Ethics* for him should not simply aim at mastering 'the concepts of attribute, substance and mode', nor just follow his system of axioms, propositions, proofs and corollaries. Rather, a scholar of Spinoza should aim to grasp what is behind the 'heavy mass' of concepts and theorems; this is for Bergson something 'subtle and very light' and it is roughly an intuition of Spinoza's sentiment of the coincidence between the act through which our mind perfectly knows the truth and the operation through which God generates it.[81] The originality of philosophy for Bergson does not lie in systems and concepts; he cited Berkeley's philosophy as an example. He argued that if one analyses it for what it explicitly says, there is very little that is new, apart perhaps from his theory of vision. His other ideas, including matter as a set of ideas, existence of minds, his nominalism and his theism, can for Bergson be found in Duns Scotus, Descartes, Hobbes and even medieval philosophers. In Bergson's view, what was really new in Berkeley's philosophy was an ineffable intuition, which he tried to convey to his Bologna audience by using the image of matter as a very thin and transparent film between man and God, through which God reveals himself. However, he continued, when metaphysicians start analysing this 'film', it becomes opaque, because terms like 'substance' gather on it just like a layer of dust.[82] For him, if one grasps Berkeley's fundamental intuition, all his other ideas will reveal themselves to be quite novel and quite different from previous formulations. The analysis of what he described as the 'external' part of philosophy for him cannot capture the real meaning of a philosophical doctrine, because philosophers themselves do not 'start from pre-existent ideas'. He thought that Spinoza did not start from the concept of substance and act, nor from his geometrical method, just as Berkeley did not start from idealism and nominalism. If pre-existing ideas seem to be present in a philosopher's work, for him this is because that is the way in which he can express and communicate his intuition, but his ideas, informed by his novel intuition, will in effect have a new sense and a new life.[83]

Let us consider the consequences of Bergson's pronouncements for the history of philosophy in turn. First of all, Bergson considered individual philosophers as the absolute authors of their work. Their use of the language, methods and ideas of their times was not what for him made their philosophies original. In his view, what the reader of their works should aim to grasp is ultimately independent of the milieu in which they lived. In a perhaps more moderate form, this was the traditional view of philosophy. In general, mainstream historians of philosophy regarded philosophical texts as the product of the individual genius and at the same time as potentially

80 Bergson, 'L'intuition philosophique', p. 176.

81 Ibid., p. 179.

82 Ibid., p. 184.

83 Ibid., p. 186.

expressing timeless truths. Indeed, for them precisely because philosophical ideas are not affected, in a deep sense, by historical contingencies, individual philosophers could communicate across centuries. The little attention that historians of philosophy paid to the historical, social, economic and cultural circumstances in which works of philosophy had appeared seems to confirm that most of them shared his view. Many, with Bréhier, regarded history, 'the domain of which is temporal', as the means to find the 'atemporal', and the 'eternal present' of philosophy.[84]

Bergson expressed the independence of philosophical ideas from the historical development of the history of philosophy in a very clear-cut manner. As he explicitly said in Bologna, the interpretation of a doctrine as a moment in a development gives a continuity to the history of philosophy that it does not really have. It follows from what he said that a history of philosophy as progress is, in his expression, the result of a 'cinematographic illusion'. His rejection of laws of historical development in the history of philosophy is clearly consistent with his view of philosophical ideas as the product of individual intuition, and with his conviction that the it was possible for the modern philosopher to grasp and indeed share, at least to a certain extent, past philosophers' intuitions. His criticism of the systems of historical development, either in the Hegelian or Comtian mould, was shared by many historians of philosophy, as I have discussed above.

Although there seemed to be almost a general agreement, among historians of philosophy, in rejecting the *a priori* philosophies of history, there was still a possibility of seeing the history of philosophy as an empirically observed development; in other words, could the past of philosophy be described as progress, although not a necessary progress? Following Bergson, the reconstruction of the history of philosophy as progressive would be an illusion as much as an *a priori* progressive model. Many historians of philosophy agreed with him, and, although their histories of philosophy may shows the connection between earlier and later philosophers, they would not have agreed that there was more truth in Aquinas's philosophy than in Plato's because the former chronologically followed the latter. The contrast between the inapplicability of the idea of progress of philosophy, and the progressive character of the history of science, was often employed by historians of philosophy, as Robin did in his talk. However, Bergson appeared to downplay the 'heavy mass' of concepts of which doctrines are formed in favour of intuition. Was this acceptable to all historians of philosophy? Moreover, did he go too far for them in his neglect of historical circumstances and textual analysis?

The History of Philosophy as Heritage and Genre

An important part of the historians of philosophy's work was to preserve a philosophical tradition, precisely because they believed in its continued relevance. They could think, as Robin did, that old scientific texts could hardly satisfy more than one's scholarly curiosity, but they attached great value to the analysis of texts by philosophers such as Plato, Descartes and Kant. Historians of philosophy preserved

84 Bréhier, *La philosophie et son passé*, p. 44.

a heritage by producing new annotated editions, exegesis and comments of classic works, and by explaining these works to generations of students. They kept this heritage alive by proposing new interpretations and new ways of employing past philosophers' ideas in order to discuss new problems. This work of conserving the heritage, however, did not find solid support in the pronouncements or practices of philosophers such as Bergson. His view of how to read past texts was very consistent with his theory of philosophical knowledge as intuition, that is as non-discursive. In Bologna, he dismissed, from a philosophical point of view, the philological and historical analysis of texts, and urged historians of philosophy to connect and empathize with the philosopher under study. His intuitive connection with past philosophers in fact devalued not only philology, but also the textual exegesis and philosophical analysis that were crucial parts of the historians of philosophy's work.

When writing history of philosophy works, historians of philosophy as a rule did analyse the 'heavy mass' of concepts presented in the text under study, and readers would have turned to their works to have that mass disentangled and clarified. As far as the history of ancient philosophy is concerned, the philological apparatus played a role that could not be overlooked. Someone who had heard Robin almost dismiss the philological and historical analysis and emphasize the importance of the more properly philosophical reading of a text, may have been surprised at the nature of his works. In his introductions to his own translations of Plato's *Symposium*, *Phaedrus* and *Phaedo*, he discussed questions such as the dating of the works and their authenticity, and analysed each section of the works separately, explaining terms, concepts and content.[85] His monographs on Plato and Aristotle are in a similar vein: after covering biography and classification of works, he examined these philosophers' works in a very systematic and thorough way.[86] The same analytical approach is to be found in his works on aspects of Plato's and Aristotle's thought.[87]

It was not the case that Robin did not agree with Bergson about philosophy being the creation of the individual philosopher, and indeed his books are mainly on individual philosophers. Rather, what for Bergson was the surface enveloping the real essence of philosophy, for Robin, more traditionally, was philosophy itself: the questions, systems, concepts and structures of reasoning. The questions posed by a philosophical text, however ancient, were for him still relevant to the modern reader, and so were its answers.[88] The 'mass of concepts' of a philosophical text that

85 Léon Robin, 'Notice', in Platon, *Banquet* (Paris: Les Belles Lettres, 1992 [1929]), Léon Robin, 'Notice', in Platon, *Phèdre* (Paris: Les Belles Lettres, 1985 [1933]), Léon Robin, 'Notice', in Platon, *Phédon* (Paris: Les Belles Lettres, 1983 [1926]).

86 Léon Robin, *Aristote* (Paris: Presses Universitaires de France, 1944); Léon Robin, *Platon* (Paris: Alcan, 1935).

87 Léon Robin, *La théorie platonicienne de l'amour* (Paris: Alcan, 1933 [1908]); Léon Robin, *La théorie platonicienne des idées et des nombres. Etude historique et critique* (Paris: Alcan, 1908). These are Robin's doctoral theses. *La théorie platonicienne de l'amour* was republished in 1933 with no changes, see Robin, *La théorie platonicienne de l'amour*, 'Avant-propos'.

88 Robin's lectures at the Sorbonne in 1932–33 were about the question of the relationship of being and knowledge; he analysed this issue in Plato, but it is clear that the problem for him

Bergson dismissed as a layer of dust covering the philosopher's true intuition was, for most historians of philosophy, the very object of their study. What, however, Robin had in common with Bergson and other philosophers was that he regarded his work as a direct engagement with philosophical texts.

Despite these profound differences in ways of conceiving the relationship with past texts, and in what it means to read a text philosophically, the idea that the truly philosophical approach to the history of philosophy is to read past texts independently of the culture in which they emerged was not only mainstream, but also enduring. Martial Gueroult has judged as truly philosophical only the work of the historians of philosophy who adopted this anti-historicist approach. He pointed out that, in the first decades of the twentieth century, two very different schools within the history of philosophy developed in France: the scholars belonging to one of them tended to produce monographs, and were interested in systems, while those belonging to the other focused on the study of long periods and philosophical movements, and on mutations of ideas as well as trends. For Gueroult, the focus on individual texts and systems develops the philosophical aspect of philosophy, with its 'affirmation of the absolute and unique, which is the hallmark of all philosophical doctrines', while the latter responds to the historical aspect of the history of philosophy.[89] Needless to say, for Gueroult, as for Bergson, Bréhier, Robin and many others, only the studies that deal with what is 'absolute and unique' are really philosophical, while those dealing with development of ideas through time are historical. And yet, challenges to this view did not all belong to the past, such as nineteenth-century systems, nor did they all come from scholars who intended to further new disciplines, such as the social scientists and the historians of science. In fact, they also came from within the ranks of the historians of philosophy: notably, Léon Brunschvicg, who attached philosophical value precisely to the study of historical development.

It is tempting to seek to establish, as Gueroult does, a correspondence between the format of the historians of philosophy's works and their historiographical ideas. It is not always easy, however, to do so, not least because the type of works that they wrote not only depended upon their convictions, but also on what was expected from academics, what was considered of value in the academic community, or indeed tacitly required in order to progress in one's a career. We should not expect dramatic departures from the formats and standards of the history of philosophy, even on the part of those scholars who did challenged mainstream ways to approach the history of philosophy. Even philosophers more committed to the study of the ideas of individual philosophers, seen as irreducible to any historical account, would write general histories of philosophy. A general history of philosophy was a grand-scale work that afforded fame, and might represent the crown of one's career. Sometimes, they were invited to write on a certain subject; for instance Berr asked Robin to write a general book on the history of Greek thought for the his series 'History of

---

was relevant to modern philosophy: Léon Robin, *Les rapports de l'être et de la connaissance d'après Platon. Cours professé en Sorbonne pendant l'année scolaire 1932–33, publié par Pierre-Maxime Schuhl* (Paris: Presses Universitaires de France, 1957).

89 Martial Gueroult, 'The History of Philosophy as a Philosophical Problem', *The Monist* 53 (1969): 563–87 (582).

Humanity', although, as we shall see, Robin did not really approved of Berr's grand idea of 'historical synthesis' that animated the series.

On the other hand, historians of philosophy committed to the idea of development of the history of thought as Brunschvicg was, did not spurn the historian of philosophy's craft of exegesis of texts, and the study of a philosopher's body of work.[90] Monographs about individual philosophers of course were standard works in the history of philosophy, and most of these historians of philosophy wrote them. Works focused on 'French philosophy' may well have been prompted either by request or by philosophical and political intents. As an example of the first group, it is worth mentioning André Lalande's (and occasionally Abel Rey's) annual correspondence on French philosophy to the *Philosophical Review* from 1905 to 1947. Similarly, Lévy-Bruhl's volume on modern French philosophy was published directly in English.[91] Works focused on the specificity of French philosophy, especially when published close to the First World War appeared to have political intentions, although there are no reasons to doubt their authors' sincerity. In 1915, Bergson presented France as the place where all modern philosophy began: modern rationalism with Descartes, philosophy of sentiment with Pascal, social philosophy with the Enlightenment philosophers, evolutionism with Lamarck, and scientific philosophy with Claude Bernard and Comte.[92] Despite his ultimately individualistic views of philosophy, Bergson did not hesitate to sketch the general character of French philosophy, comprising clarity and use of every-day language, closeness to the positive sciences, a tendency to introspection and psychological research, and a profound suspicion of systems and dogmas, including Kant's and Hegel's.[93] These traits for Bergson made French philosophy close to English and American philosophical traditions, but far removed from German philosophy. Many other philosophers, also within the cohort of historians of philosophy, agreed with Bergson. Delbos wrote an entire volume dedicated to French philosophy, published in 1919. He also found several national

---

90 See, for instance, Léon Brunschvicg and Pierre Boutroux (eds), *Œuvres de Blaise Pascal, publiées suivant l'ordre chronologique avec documents complémentaires, introductions et notes*, vols 1–3 (Paris: Hachette, 1908); Léon Brunschvicg, Pierre Boutroux, and Félix Gazier (eds), *Œuvres de Blaise Pascal, publiées suivant l'ordre chronologique avec documents complémentaires, introductions et notes*, vols 4-6 (Paris: Hachette, 1914); Léon Brunschvicg, 'La logique de Spinoza', *Revue de métaphysique et de morale* 1 (1893): 453–67. This article was going to become chapter 2 of Léon Brunschvicg, *Spinoza* (Paris: Alcan, 1894). He continued to study Spinoza, see Léon Brunschvicg, 'Spinoza et ses contemporains', *Revue de métaphysique et de morale* 13 (1905): 637–705, Léon Brunschvicg, 'Spinoza et ses contemporains (suite)', *Revue de métaphysique et de morale* 14 (1906): 35–82; Léon Brunschvicg, 'Spinoza et ses contemporains (suite et fin)', *Revue de métaphysique et de morale* 14 (1906): 691–732. These three articles were then republished, with some modifications, together with his previous book on Spinoza to produce a new monograph: Léon Brunschvicg, *Spinoza et ses contemporains* (Paris: Alcan, 1923).

91 Lucien Lévy-Bruhl, *History of Modern Philosophy in France* (London: Kegan Paul, Trench, Trübner & Co., 1899).

92 Bergson, 'La philosophie française (Tableau récapitulatif destiné à l'Exposition de San Francisco)'.

93 Ibid., p. 251–2.

characteristics of French philosophy, including clarity and interest in psychology and ethics.[94]

However, Gueroult is not wrong in pointing out that the historians of philosophy's different approaches to their material were reflected in the format of their work. Although most of them practised the standard genres of the history of philosophy, including exegesis of texts, works on individual philosophers, on periods or schools, and general histories of philosophy, the relative weight of these works in their overall production was different. In Brunschvicg's body of work, histories of philosophy, sciences and concepts dominate not just in quantity but in the philosophical investment that their author made in them. At the other end of the spectrum, we find historians of philosophy whose more important works were on individual philosophers and exegesis of texts. For a historian of philosophy such as Robin, studies of the works of Plato, Aristotle or other classic authors[95] were also aimed to preserve a cultural tradition and a heritage. The great works of philosophy were for Robin and for many of his colleagues classics in the literal meaning of the term: rather than being superseded by new works and theories, they constituted the exemplary texts that always deserved to be re-read and re-interpreted. The conception of history as heritage was deeply embedded in the history of philosophy. Among the philosophers, historians of philosophy were those who kept past texts and the practice of their interpretation alive. They were at the same time the guardians of the tradition, but also those who prevented the tradition from becoming 'mere' historical record. Other conceptions of history, however, challenged their view: on the one hand, the conception of history as development and progress, and the view according to which past ideas are the product of different ways of thinking, and therefore cannot be directly relevant to present debates.

---

94 Victor Delbos, *La philosophie française* (Paris: Plon, 1919), pp. 3, 13.

95 See Léon Robin, *Pyrrhon et le scepticisme grec* (Paris: Presses Universitaires de France, 1944).

Chapter 3

The Meaning and Uses of History: Challenges to the History of Philosophy

History as Heritage versus History as Evolution: The Challenge of the Social Sciences

For mainstream historians of philosophy, the 'history' of philosophy was a heritage not only to be preserved, but also to be kept relevant to modern times. Their conviction that past philosophical texts were still highly relevant to the modern reader was based on the assumption that philosophical writings investigate timeless questions. If old texts at first sight looked unfamiliar to the modern reader, it was the task of the historian of philosophy to reveal their enduring significance underlying the alien language and style. Historians of philosophy realized that their discipline, as they understood it, was undermined by those philosophies of history according to which ideas are relative to the time and culture in which they emerge, or receive their meaning from their position within a grand narrative. In this regard we have seen Delbos and Bréhier's opposition to Comte's and Hegel's philosophies of history, which was widely shared by as diverse philosophers as Bergson, Brunschvicg, Gilson and Robin. Obviously, these two nineteenth-century philosophers could not represent a threat in person. Hegel's reception in France, relatively marginal before Wahl's and Hyppolite's works,[1] focused, as briefly mentioned with regard to Wahl, on aspects of his philosophy well removed from the laws of historical development.

Comte, however, was another matter. The discipline to which he had given a name, sociology, was part of the philosophy curriculum. The founding father of modern sociology, Emile Durkheim, had extensively criticized Comte, but could his own conception of historical development be in harmony with the history of philosophy as heritage? There certainly were great difficulties. Durkheim regarded human history as a development from simpler to more complex forms of society. For him, the study of the past has value in relation to the present, and it should be

---

1 See Wahl's seminal work, Wahl, *Le malheur de la conscience dans la philosophie de Hegel*, and his previous article Jean Wahl, 'Commentaire d'un passage de la phénoménologie de l'esprit de Hegel', *Revue de métaphysique et de morale* 34 (1927): 441–71; see also Hyppolite's articles: Jean Hyppolite, 'Les travaux de jeunesse de Hegel d'après des ouvrages récents', *Revue de métaphysique et de morale* 42 (1935): 399–426; Jean Hyppolite, 'Les travaux de jeunesse de Hegel d'après des ouvrages récents (suite et fin)', *Revue de métaphysique et de morale* 42 (1935): 549–78; and Jean Hyppolite, *Introduction à la philosophie de l'histoire de Hegel* (Paris: Rivière, 1948). Alexandre Kojève's seminars on Hegel at the Ecole pratique des hautes études were also of great importance; Alexandre Koyré, whose work will be discussed in Chapter 5, attended them regularly.

carried out in order to understand how the present has come into being.[2] By contrast, historians of philosophy regarded their texts as timeless sources of truth and wisdom, capable of engaging modern philosophers in a dialogue.

Equally threatening for mainstream historians of philosophy were those theories that transferred the models of evolution that Jean-Baptiste Lamarck and Charles Darwin had proposed for organisms to intellectual forms and ideas. It was at one of the *RMM* international conferences, held in Geneva in 1904, that Emile Boutroux argued that evolutionism, in its different versions, undermined the history of philosophy as part of philosophy. In his talk, Boutroux in particular referred to the transfer of the theory of evolution from biology to philosophy. First of all, he explained, from evolutionistic theories follow that contemporary questions are not the same as those posed in the past. Although evolutionists think that the present is the result of history, he continued, they either think that history itself has already operated the necessary selections of doctrines that have survived, and therefore there is no need to study others, or alternatively think that the past is worth of interest, but only as a way of understanding the present.

Boutroux did not mention any French scholars who had argued for these positions; the only living scholar he did mention was the Italian positivist Roberto Ardigò,[3] whose view, as presented by Boutroux, was that the history of philosophy, logically ordered according to evolutionistic principles, constituted the evidence of the 'methodical labour' of the human mind and the key of philosophy.[4] The author of *De la contingence des lois de la nature* could not accept laws of evolution for philosophy. Faced with the evolutionistic view, Boutroux retorted that in the history of philosophy the past is never superseded once and for all, and provided many examples of philosophers 'going back' to old conceptions, including colleagues who

---

2 For a concise, but comprehensive and clear discussion of Durkheim's conception of history and his method of historical analysis, see Warren Schmaus, *Durkheim's Philosophy of Science and Sociology of Knowledge: Creating and Intellectual Niche* (Chicago and London: The University of Chicago Press, 1994), pp. 113–7. An interesting early example of the difficulties between Durkheim and the *RMM* philosophers is his proposal for a national philosophy syllabi in schools, and teachers' training programmes and examinations (Emile Durkheim, 'L'enseignement philosophique et l'agrégation de philosophie', *Revue philosophique* 39 [1895]: 121–47). Among other proposals, Durkheim suggested that the study of the historical process whereby the current conceptions and practices have been formed should be part of philosophy. Unlike historians of philosophy, however, he regarded the history of philosophy just as a small part of humankind's past, without any special status (Emile Durkheim, 'L'enseignement philosophique et l'agrégation de philosophie': 142). The *RMM* philosophers immediately issued a reply in which they declared that it was 'dangerous to have Durkheim as an ally', for he wanted to reduce all philosophy to the logic of the social sciences. (N.D.L.R., 'Enseignement', *Revue de métaphysique et de morale* 3 [1895]: 231–3 [232]). For a discussion of this controversy see Barberis, 'Moral Education for the Elite of Democracy'.

3 The publications of Roberto Ardigò (1828–1920) include Roberto Ardigò, *La psicologia come scienza positiva* (Mantova: Viviano Guastalla, 1870) and Roberto Ardigò, *La morale dei positivisti* (Milano: Natale Battezzati, 1889).

4 Boutroux, 'Rôle de l'histoire de la philosophie dans l'étude de la philosophie', pp. 50–51.

called for a return to Leibniz or Kant. Philosophical ideas are never really lapsed for Boutroux; Kant might have declared metaphysics dead, he argued, but this did not prevent the 'splendid development' of German metaphysics at the beginning of the nineteenth century.[5] For Boutroux, the 'individual and contingent' are inseparable from philosophy and its history, which cannot be ordered according to any law or plan.[6]

The models of intellectual history that were based on theories of progress or evolution presented two problems for mainstream historians of philosophy. The first was that past philosophical texts inevitably were judged less sophisticated than modern ones, indeed superseded by more modern theories. The other was that past texts became documents of earlier stages of human mental and social development. Robin, Rivaud, Gilson and many others did not regard their ancient or medieval texts either as primitive forms of present-day philosophies, or as evidence of past ways of thinking. Rather, they studied them because they thought that they raised still relevant questions, and that they expressed truths that had not been systematically overcome.

It was not only the social scientists or positivist philosophers across the Alps who evaluated philosophical texts differently. In fact, a very influential academic, the Professor of History of Modern Philosophy, Lucien Lévy-Bruhl, did read texts as evidence of the functioning of the mind. Traditional historians of philosophy did not read Plato to find out the worldview of fourth-century-BCE Athenians, nor did they study Descartes in order to analyse the way of reasoning of seventeenth-century French people. Lévy-Bruhl, however, seemed to do precisely that: he aimed to study the mind, and analysed texts in order to understand it. Since his aim was to study the mind, there was no reason to limit himself to philosophical texts, and indeed he moved on to ethnological reports.

The next Sorbonne Professor of History of Modern Philosophy, Léon Brunschvicg, opposed both laws of historical development and positivism, and, as discussed, was in the 'inner' circle of the *RMM* group. This might seem to suggest that he sided with mainstream historians of philosophy. However, the content of his books appears to be about the evolution of the mind, and his approach to texts looks close to that of Lévy-Bruhl, in that he saw them as sources of information about the capabilities of the mind. I shall discuss his approach in the last section of this chapter, for he is in my view the most important link between the history of philosophy and the history of science conceived as the study of the mind. First, though, it is necessary to discuss Lévy-Bruhl's view of the mind, for it is of great importance not only for Brunschvicg, but for the historians of science whom we will encounter in the following chapter.

5 Ibid., p. 53.

6 Ibid., p. 57.

The Mind and Mentalities: Lucien Lévy-Bruhl

Lévy-Bruhl: A Non-Mainstream Historian of Philosophy

Lévy-Bruhl's early publications appeared to be in line with what could be expected of a mainstream philosopher who was orienting his interests towards the history of philosophy, publishing as he did monographs on Jacobi, Comte and modern philosophy in France.[7] He had an interest in German thought; indeed, in addition to his monograph on Jacobi, he also published a volume on Germany since Leibniz.[8] Jean Duvignaud has noticed that in these two books Lévy-Bruhl presents an image of Germany that is not that of 'Germany as France's enemy', which some of his contemporaries promoted.[9] Lévy-Bruhl did not refrain, when still a *lycée* teacher, from delivering a lecture on Hegel at the Académie des sciences morales et politiques when his philosophy was perhaps still seen by many as the 'philosophy of Prussian brutality', as the philosopher and jurist Emile Beaussire had put it after 1870 in the same lecture theatre.[10] There was nothing exceptional about Lévy-Bruhl's interest in German thought, which he indeed shared with such illustrious philosophers as Boutroux and Delbos. However, he did not counterpoise French and German traditions as did Bergson, for instance, as mentioned in the previous chapter. Moreover, it is interesting that, while historians of philosophy attacked Hegel's and Comte's systems, Lévy-Bruhl not only lectured on Hegel, but also on Comte.[11] Indeed, in 1900, when he was already *maître de conférences* at the Sorbonne, he published a monograph on the latter philosopher, which at first sight appeared to be in line with the traditional approach of historians of philosophy.[12] Comte was of course a modern philosopher, and, for many historians of philosophy, even a controversial one; in

7 Lévy-Bruhl, *History of Modern Philosophy in France*; Lucien Lévy-Bruhl, *La philosophie d'Auguste Comte* (Paris: Alcan, 1900); Lucien Lévy-Bruhl, *La philosophie de Jacobi* (Paris: Alcan, 1894).

8 Lucien Lévy-Bruhl, *L'Allemagne depuis Leibniz. Essai sur le développement de la conscience nationale en Allemagne, 1700–1848* (Paris: Hachette, 1890). Before this work, Lévy-Bruhl had published his doctoral theses: Lucien Lévy-Bruhl, *L'idée de responsabilité* (Paris: Hachette, 1884); Lucien Lévy-Bruhl, *Quid de Deo Seneca senserit* (Paris: Hachette, 1884). For a full bibliography of Lévy-Bruhl's works, see the special issue of the *Revue philosophique* dedicated to him: vol. 114 (1989).

9 Jean Duvignaud, *Le langage perdu. Essai sur la différence anthropologique* (Paris: Presses Universitaires de France, 1973), p. 95.

10 Bernard Bourgeois, 'Lévy-Bruhl et Hegel', *Revue philosophique* 114 (1989): 449–51 (449).

11 Marcel Mauss, 'Lévy-Bruhl sociologue', *Revue philosophique* 64 (1939): 251–3 (251); Georges Dumas, 'Lévy-Bruhl, La philosophie d'Auguste Comte', *Revue philosophique* 25 (1900): 396–407 (369).

12 Lévy-Bruhl, *La philosophie d'Auguste Comte*. For Comte's centenary, he published an article in the *Revue des deux mondes*: Lucien Lévy-Bruhl, 'Le centenaire d'Auguste Comte', *Revue des deux mondes* 145 (1898): 394–424. He had also published Comte's correspondence with John Stuart Mill, after finding Mill's letters in Comte's papers: Lucien Lévy-Bruhl (ed.), *Correspondance de John Stuart Mill et d'Auguste Comte* (Paris: Alcan, 1899).

other words, he was not on the same level as Pascal and Plato, the philosophers on whom Boutroux and Milhaud published books in the same year.[13]

It is in the first chapter of his monograph on Comte, however, that the attentive reader would have detected a departure from the approach that was standard in teaching and mainstream in research. Lévy-Bruhl rejected the method of reading texts directly, without situating them historically. His view of how philosophies emerge goes further than a very sympathetic reviewer like Georges Dumas[14] gave him credit for. Dumas remarked that Lévy-Bruhl discarded the method of adopting the 'mental attitude' of the author under study and had analysed his work 'objectively'.[15] Dumas's remark seems to contradict Gilson's view, reported in Chapter 1, that Lévy-Bruhl's books on Jacobi and Comte taught how to present philosophers' thought from their own points of view. However, Dumas's and Gilson's different views of Lévy-Bruhl's method have something in common: they both emphasize that Lévy-Bruhl did not seek to grasp the fundamental intuition at the core of a philosophy, as we have seen Bergson propose. Similarly, both Dumas and Gilson implicitly agreed that Lévy-Bruhl did not share the methodology of Robin, Wahl, Bréhier and Boutroux, who regarded the reading of classic texts as affording the possibility for the readers to find a mind similar to their own,[16] or to make the reader's mind 'vibrate in sympathy'.[17]

Lévy-Bruhl, however, did something more controversial than just analysing Comte's philosophy objectively. In his introduction to *La philosophie d'Auguste Comte*, he argued that it is not possible to study a philosophical system in isolation, for 'the influence of the religious, political, economical, intellectual phenomena, in a word of the contemporary milieu upon this system is as indisputable as its own influence upon the milieu'. Indeed, he argued in favour of a causal link between social events and philosophical doctrines, when he claimed that, without the French revolution, 'neither the theory of progress, nor consequently social science, nor consequently again positive philosophy would have been possible'. For him the French revolution posed the questions that all 'thinking people' had to address, including what principles should guide the construction of a new social order.[18] Here Lévy-Bruhl's view of philosophical texts was at odds with those of most historians of philosophy, for he claimed that philosophical ideas are not timeless, but that in fact their emergence causally depends on particular historical and social situations.

---

13 Emile Boutroux, *Pascal* (Paris: Hachette, 1900); Gaston Milhaud, *Les philosophes-géomètres de la Grèce. Platon et ses prédécesseurs* (Paris: Alcan, 1900).

14 Georges Dumas (1866–1946), philosophy *agrégé*, obtained doctorates both in humanities and medicine. He started his career as a philosophy teacher in secondary schools, and eventually became professor of experimental psychology (1913) and then of pathological psychology (1921) at the University of Paris. At the time of the Lévy-Bruhl review, he was the director of the psychology laboratory of the Faculty of Medicine in Paris. He edited the classic *Traité de psychologie*: Georges Dumas (ed.), *Traité de psychologie* (Paris: Alcan, 1923–24).

15 Dumas, 'Lévy-Bruhl, La philosophie d'Auguste Comte', p. 397.

16 Boutroux, 'Rôle de l'histoire de la philosophie dans l'étude de la philosophie', p. 58.

17 Bréhier, 'La philosophie et son passé', p. 44.

18 Lévy-Bruhl, 'La philosophie d'Auguste Comte', pp. 1–2.

In his study of Comte, Lévy-Bruhl did not really attempt to show the other 'parts' on which Comte's philosophy was at the same time dependent and had an influence. However, his subsequent books represent, in my view, his attempt to analyse the extra-textual parts of philospphy. His next monograph, *La morale et la science des mœurs*,[19] is rightly regarded as the first work in which he adopted the sociological approach that was going to characterize his subsequent volumes on 'primitive mentality'.[20] It is in *La morale* that Lévy-Bruhl more fully conceived of a method for investigating human ideas and behaviours that would overcome the abstract and *a priori* form of philosophical discussions. Its continuity with *La philosophie d'Auguste Comte*, however, is strong. It is not just that, after studying Comte, Lévy-Bruhl decided to adopt a sociological and positivist approach to the study of morals. He also applied the view expressed in his previous book, namely that philosophical ideas are part of a larger social and historical reality. If philosophical ideas are part of a 'whole', the other parts must be crucial for the comprehension of those ideas and more generally of human beings.[21] In *La morale*, as a good historian of philosophy, he examined the ethical theories of Kant, Locke and Leibniz, but found them inadequate for a full understanding of human morality. He rejected the conception of ethics as an *a priori* set of theories, outside time and social circumstances. Just as for him there is no philosophical text outside time and place, so there is no human behaviour that can be studied in the abstract. Rather, he claimed that:

> Instead of speculating on man as a being naturally moral, we must discover how the whole mass of prescriptions, obligations and interdictions which constitute the ethics of a given society, is formed in relation to other series of social phenomena.[22]

La morale et la science des mœurs is the most Durkheimian and Comtian of Lévy-Bruhl's works, as Merllié remarks.[23] This is also apparent as far as the discussion about

---

19 Lucien Lévy-Bruhl, *La morale et la science des mœurs* (Paris: Alcan, 1903).

20 Dominique Merllié, 'Présentation. Le cas Lévy-Bruhl', *Revue philosophique* 114 (1989): 419–48 (433); Cazeneuve, *Lucien Lévy-Bruhl: sa vie, son œuvre*, p. 17; Patrizia Di Palma, *Lucien Lévy-Bruhl: dalla scienza dei costumi all'antropologia* (Lucca: Pacini, 1983). Di Palma's book is entirely aimed at showing the link between *La morale et la science des mœurs* and his works on mentalities. She also argues that Lévy-Bruhl's book on Jacobi already showed his interest in not entirely rational *formae mentis* (p. 11).

21 Duvignaud, who cites Georges Davy on this point, argues that there is a strong continuity between Lévy-Bruhl's work on Comte and his subsequent anthropological work. For Duvignaud, in his work on Comte, Lévy-Bruhl realized that a system of thought could be studied objectively, rather than through the subjectivity of the commentator. This was the method that he subsequently employed in his anthropological work: Duvignaud, *Le language perdu*, pp. 95–7. Duvignaud's comment is reminiscent of Dumas's.

22 Lucien Lévy-Bruhl, *Ethics and Moral Science* (trans. Elisabeth Lee; London: Archibald Constable, 1905 [1903]), p. 166.

23 Merllié, 'Présentation. Le cas Lévy-Bruhl': 433. Durkheim wrote a positive review of *La morale* in the *Année sociologique*, saying that Lévy-Bruhl's book, which displayed 'rare dialectic rigour', discussed the science of moral facts, and that the existence of this science was at the basis of all the *Année sociologique* group's work. Emile Durkheim, 'Lévy-Bruhl, La morale et la sciences des mœurs', *Année sociologique* 7 (1902–1903): 380–84.

historical models and conceptions of history is concerned. In this book, Lévy-Bruhl claimed that 'practical ethics' exhibits 'a slow but almost uninterrupted evolution'. In his later works, Lévy-Bruhl abandoned the Durkheimian view of evolution of ideas and practices. His subsequent, and more famous, books studied the 'mentalities' of traditional societies in Africa, the Americas, Asia and Oceania. He did not see his new subject as a departure from his philosophical concerns: as he wrote to Evans Pritchard in 1934, he did not intend to contribute to anthropology or ethnology; rather, he aimed to study human nature by using the data provided by ethnology. He added that his training had been in philosophy, rather than anthropology, and his cultural roots were in the works of Spinoza and Hume, rather than those of the anthropologists.[24] Indeed, his own reconstruction of how his interests shifted is that of a seamless evolution. He said that, while the French sociological school was developing thanks to Durkheim, he was occupied with his work in the history of philosophy. One day, however, he received a book in three volumes of a Chinese historian, translated into French. He was struck by the different way in which this historian connected his ideas, and started wondering whether that Chinese author followed a different logic from his own, and that of western people in general. A philosophical problem was before him: is human nature always and everywhere identical, as philosophers from Fontenelle and Hume to Comte had maintained?[25] This was the question that he set out to answer with his books on 'primitive mentality'.

Beyond Evolution versus Timeless Texts: Primitive and Modern Mentalities

Despite the continuity between Lévy-Bruhl's early works and his books on mentalities, the latter do mark a shift in his conception of historical development. His works on 'primitive mentality' increasingly focused on the differences between ways of thinking rather than on the evolution of one mentality from another.[26] In very general terms, he held that the so-called 'primitives', that is to say people belonging to traditional and non-literate societies, thought differently from modern people. 'Primitives', according to contemporary reports, not only had not produced a science or type of knowledge similar to western people, but also seemed systematically to explain phenomena in a very different way, and looked at objects differently. Lévy-Bruhl discarded the missionaries' explanation that the thought of those peoples did not step beyond the limited realm of their senses, in other words that 'primitives' were neither capable of abstract thought, nor were interested in what was not immediately before their eyes. He argued that children from those societies were just

24 Quoted in Duvignaud, *Le langage perdu*, p. 126.

25 Lévy-Bruhl et al., 'La mentalité primitive. Séance du 15 février 1923', pp. 20–22.

26 Lévy-Bruhl took up the term 'mentality' in *La mentalité primitive* (1922); the term already appeared in 'anticipation' of the content of this famous volume; see Lucien Lévy-Bruhl, 'La mentalité primitive et les médicins européens', *Revue de Paris* 16 (1921): 806–826. He claimed that he did not use it in *Le fonctions mentales dans les sociétés inférieures* (1910) because it had not yet entered common use. He admitted that it was rather vague, introduced and made popular by journalists, but had then become current; see Lucien Lévy-Bruhl, *La mentalité primitive: The Herbert Spencer Lecture, delivered at Oxford 29 May 1931* (Oxford: Clarendon Press, 1931), p. 7.

as capable of learning what missionaries taught them as French or English children were; it followed that these differences in thinking could not be put down to a lesser ability.[27]

Lévy-Bruhl departed from the more common sociological model of evolution when he stated that it was incorrect to regard their different way of thinking simply as a 'rudimentary form' of our mental activity, or as an infantile form.[28] His hypothesis was that it was just different, rather than an earlier stage of modern mental activity. He believed that 'primitives' exhibit a 'prelogical' mentality, which did not mean either 'alogical or antilogical', but rather than they were not concerned with avoiding contradiction.[29] Specific to their mentality was rather the so-called law of participation. According to this law, beings can be at the same time themselves and something else; moreover, without ceasing to remain where they are, they can emit and receive mystic powers, virtues and qualities, which can be felt elsewhere. Lévy-Bruhl provided the examples of the Trumai, a Brazilian tribe, who claimed that they were aquatic animals and a nearby tribe, the Bororo, who boasted of being red parakeets.[30] Lévy-Bruhl discussed many other characteristics that according to him made 'primitive' thought different from modern thought. Among these, he indicated that 'primitives' disregard secondary causes, and explain natural phenomena with supernatural causes;[31] in other words, their causality is mystical.[32] For him, the separation between natural and supernatural, and the consequent concept of natural laws, do not exist for the 'primitives'.

This polarity between primitive and modern mentality, although it is at the core of Levy-Bruhl's theory, should not necessarily be understood in the sense that one individual, or one civilization, exclusively and permanently has either mentality. Some modern commentators have found the idea that the same individual has more than one mentality unacceptable.[33] However, this was acceptable to Lévy-

---

27 Lucien Lévy-Bruhl, *La mentalité primitive* (Paris: Alcan, 1922), pp. 12–13.

28 Ibid., p. 15; Lucien Lévy-Bruhl, *Les fonctions mentales dans les sociétés inférieures* (Paris: Alcan, 1910), p. 76.

29 Lévy-Bruhl, *Les fonctions mentales dans les sociétés inférieures*, p. 79; Lévy-Bruhl et al., 'La mentalité primitive. Séance du 15 février 1923', p. 18; Lévy-Bruhl, *La mentalité primitive: The Herbert Spencer Lecture*, p. 21.

30 Lévy-Bruhl, *Les fonctions mentales dans les sociétés inférieures*, p. 77. Mariapaola Fimiani has offered a careful presentation of Lévy-Bruhl's law of participation in: Mariapaola Fimiani, *L'arcaico e l'attuale: Lévy-Bruhl, Mauss, Foucault* (Torino: Bollati Boringhieri, 2000), pp. 41–52. Frédérick Keck has argued that Lévy-Bruhl's concept of participation has its roots in Malebranche's metaphysics, and that this link to Malebranche explains the peculiar position of Lévy-Bruhl's thought in relation to that of his contemporaries Durkheim and Bergson: Frédérick Keck, 'Causalité mentale et perception de l'invisible. Le concept de participation chez Lévy-Bruhl', *Revue philosophique* 130 (2005): 304–322. See also Francesco Saverio Nisio, 'Partecipazione come *scientia intuitiva*, Lévy-Bruhl e Spinoza', *Revue philosophique* 130 (2005): 323–33.

31 Lévy-Bruhl, *La mentalité primitive*, ch. 1.

32 Ibid., pp. 127ff.

33 See for instance G.E.R. Lloyd, *Demystifying Mentalities* (Cambridge: Cambridge University Press, 1990), p. 5. Lloyd mentions that Le Goff did allow for one individual to have several mentalities: Lloyd, *Demystifying Mentalities*, p. 5.

Bruhl, and even more so to Hélène Metzger and Gaston Bachelard, as we shall see in the following chapters. Already in *Les fonctions mentales*, Lévy-Bruhl wrote that logical and prelogical mentalities coexist in the primitives' way of thinking, while they are mutually exclusive for modern people, for, although some of their collective representations are 'prelogical', they do not have the same strength as those of primitive people'.[34] Later on, possibly stimulated by Hélène Metzger's article on him, he suggested that there is a 'primitive core' of the human mind that never disappears completely, and may be the origin of poetry, metaphysics and even scientific discoveries.[35]

On the one hand, Lévy-Bruhl broke with the idea of a timeless human mind, and timeless concerns and questions that were at the core of the historiography of the majority of historians of philosophy. On the other hand, he proposed an alternative to the historiography of evolution, favoured by the Durkheimians, according to which 'primitive' forms of thought should be regarded as previous stages of the modern way of thinking. In his review of *Les fonctions mentales*, Durkheim compared Lévy-Bruhl's view with his own as expressed in the *Les formes élémentaires de la vie religieuse.*[36] Against Lévy-Bruhl, he argued that there is no opposition between 'primitive' and modern ways of thinking, indeed, the 'higher and most recent forms' of thought are born of the 'most primitive and inferior' forms. In *Les formes*, Durkheim had set out to study 'the simplest and most primitive religions',[37] and had concluded that:

> There is no gulf between the logic of religious thought and the logic of scientific thought. Both are made up of the same essential elements, although these elements are unequally

---

34 Lévy-Bruhl, 'Les fonctions mentales dans les sociétés inférieures', p. 133.

35 Lévy-Bruhl, *La mentalité primitive: The Herbert Spencer Lecture*, pp. 26–7. Metzger's article (Hélène Metzger, 'La philosophie de Lévy-Bruhl et l'histoire des sciences', *Archeion* 12 [1930]: 15–24) will be discussed in Chapter 5. Lévy-Bruhl views evolved from his first book in 1910 to the end of his life, partly as a result of fieldwork reports that were made available, and partly as a constant reflection and re-working of his concepts (see Robin Horton, 'Lévy-Bruhl, Durkheim and the Scientific Revolution', in Robin Horton and Ruth Finnegan [eds], *Modes of Thought: Essays on Thinking in Western and Non-Western Societies* [London: Faber & Faber, 1973], pp. 249–305 [279]). The last chapter of his long reflection was his *Carnets*, a series of notes published posthumously (Lucien Lévy-Bruhl, *Carnets* [Paris: Presses Universitaires de France, 1949]). Although several critics have seen the *Carnets* as a recantation of his theory (see examples in Horton, 'Lévy-Bruhl, Durkheim and the Scientific Revolution', p. 257, n. 4), more recently Lévy-Bruhl scholars have rightly argued that no real break can be detected in his doctrine, but rather an evolution and further definition (see Horton and Finnegan [eds], *Modes of Thought*, pp. 257–8; Duvignaud, *Le langage perdu*, p. 94; Merllié, 'Présentation. Le cas Lévy-Bruhl', p. 428). In any case, the *Carnets*, published in 1949, could obviously have no impact in the fortune of the theory of primitive mentality before the war, when it took roots not only in the social sciences, but in many other disciplines, including general history and the history of science, as I shall discuss in the following chapters.

36 Emile Durkheim, *Les formes élémentaires de la vie religieuse. Le système totémique en Australie* (Paris: Alcan, 1912).

37 Emile Durkheim, *The Elementary Forms of Religious Life* (trans. Karen E. Fields; New York: The Free Press, 1995 [1912]), p. 1.

> and differently developed. What appears above all to typify the logic of religious thought is a natural taste as much for unrestrained assimilations as for clashing contrasts … As a result, it employs logical mechanisms with a certain gaucheness, but none of them are unknown to it.[38]

In a discussion of *La mentalité primitive* that took place in 1923 at the Société française de philosophie, Lévy-Bruhl again defended his position. He argued that the Durkheimians assumed that the 'mental functions' are everywhere the same. He claimed that was precisely the hypothesis that he opposed. The differences between societies for him were not only about costumes, beliefs and institutions; in fact he believed that different people exhibited different mental functions.[39] Marcel Mauss defended the Durkheimian position. He prefaced his comments by saying that he and Lévy-Bruhl were sociologists because both believed that the mind has a history and that this history cannot be written without writing at the same time the history of societies. However, he also said ironically that he was probably already able to show his gratitude to the Société for inviting him by offering it the spectacle of two sociologists who 'devour each other'.[40] He was referring to the fundamental disagreement between the Durkheimians, like himself, and Lévy-Bruhl on how to write the history of the mind. Mauss lamented that Lévy-Bruhl had lumped together peoples who were 'at different stages' of development: some, like the Australians, were primitive, the Americans and the Polynesians were in the Neolithic Age, and African and Asian societies had already overcome the Stone Age.[41] In other words, Mauss defended the idea of a development of civilization that is common to all human beings; different societies are just at different stages. Indeed, for him there was no break between primitive and scientific ways of organizing experience. He had not changed his mind since he and Durkheim had written that:

> Primitive classifications are therefore not singular or exceptional, having no analogy with those employed by more civilized peoples; on the contrary, they seem to be connected, with no break in continuity, to the first scientific classifications.[42]

38 Ibid., p. 240. For a contemporary discussion of the differeny views of Durkheim and Lévy-Bruhl, see Georges Davy, *Sociologues d'hier et d'aujourd'hui* (Paris: Alcan, 1931), ch. 5; for recent ones, see Warren Schmaus, 'Lévy-Bruhl, Durkheim, and the Positivist Roots of the Sociology of Knowledge', *Journal of the History of the Behavioral Sciences* 32/4 (1996): 424–40; and Horton, 'Lévy-Bruhl, Durkheim and the Scientific Revolution'. Merllié also has discussed the reasons why Lévy-Bruhl seemed to have hardly replied to the Durkheimians' criticism: Dominique Merllié, 'Lévy-Bruhl et Durkheim. Notes biographiques en marge d'une correspondance', *Revue philosophique* 114 (1989): 493–513.

39 Lévy-Bruhl et al., 'La mentalité primitive. Séance du 15 février 1923', p. 23.

40 Mauss in ibid., p. 24. Marcel Mauss (1872–1950), at the time of the discussion was directeur d'études at the Ecole pratique d'hautes etudes. He would be elected Professor at the Collège de France in 1930.

41 Ibid., p. 26.

42 Emile Durkheim and Marcel Mauss, 'De quelques formes primitive de classification: contribution à l'étude des représentations collectives', *L'année sociologique* 6 (1901–1902): 1–72 (66). The English translation is from Emile Durkheim and Marcel Mauss, *Primitive Classification* (trans. Rodney Needham; London: Cohen & West, 1969 [1901–1902]), p. 81.

Gustave Belot supported Mauss's view, saying that it explained the continuity of primitive and modern mentalities, whereas, if Lévy-Bruhl was right, we should observe mental revolutions, which, however, history does not exhibit.[43] Indeed, Lévy-Bruhl regarded 'primitive' mentality as different from modern mentality, rather than simply an early stage of the latter. However, to some Durkheimians, Lévy-Bruhl's doctrine seemed a version of Comte's. Indeed, Paul Fauconnet,[44] said that Lévy-Bruhl's primitive mentality looked like Comte's theological stage. He probably wanted to recall the historian of philosophy Lévy-Bruhl's work on Comte. Although as a rule not many philosophers or sociologists liked Lévy-Bruhl's rejection of the universality of the mind, some voices spoke in his favour on that occasion. One of them came from Henri Piéron, a psychologist especially interested in animal psychology, who defended Lévy-Bruhl. He believed that his opponents, who assumed that the human mind was always and everywhere the same, interpreted 'primitive' thought by analogy with their own, falling into what he called 'Euromorphism' [*européomorphisme*]. The unity or disunity[45] of the human mind was something to be proved empirically, he argued, rather than assumed, and Lévy-Bruhl had proved his scientific credentials by not assuming either position, but rather by reaching his conclusion through research.[46]

Unsurprisingly, the philosophers contradicted Lévy-Bruhl's thesis on the grounds of principles and logic rather than empirical results. One of their arguments, which reminds the modern reader of the objections to Thomas Kuhn's thesis of incommensurability, was that, if primitive mentality was really different from modern mentality, then we would not be able to understand it. In this particular meeting, Dominique Parodi aired this objection. For an animal psychologist like Piéron, this was not a strong objection: he admitted that if one wants to use introspection, differences in ways of thinking could be insurmountable obstacles, but on the other

---

43 Belot in Lévy-Bruhl et al., 'La mentalité primitive. Séance du 15 février 1923', p. 31.

44 At the time of that session, Fauconnet was *maître de conferences* in sociology and education at the Sorbonne. His career was rather standard for a French sociologist of the time: he taught philosophy in several *lycées*, then lectured at the University of Toulouse in social philosophy; eventually, he was appointed Professor of Sociology at the Sorbonne (1926). He wrote with Mauss the article 'Sociologie' in the *Grand encyclopédie*, vol. XXX (1902); see Guigue, *La Faculté des Lettres de l'Université de Paris*, p. 251; Charle, *Les professeurs de la Faculté des Lettres de Paris*, pp. 79-80.

45 Piéron employed the term 'unity' of the mind in the sense of 'universality', or the mind being the same in all places and times.

46 Piéron in Lévy-Bruhl et al., 'La mentalité primitive. Séance du 15 février 1923', pp. 43–4. Henri Piéron (1881–1964) was director of the laboratory of physical psychology at the Sorbonne, and would be elected Professor of Physiology of Sensations at the Collège de France later that year (1923). Like his fellow psychologists Charles Blondel and Henri Wallon, he was a *lycée* Louis-le-Grand alumnus: see Charle and Telkès, *Les professeurs du Collège de France*, pp. 203–205. His double training, in philosophy and medicine, was standard among experimental psychologists. For another psychologist's view of Lévy-Bruhl's *La mentalité primitive*, see Henri Wallon, 'La mentalité primitive et celle de l'enfant', *Revue philosophique* 53 (1928): 82–105.

hand in his view differences did not prevent 'external' observation of the kind that he carried out on animals.[47] In other words, Piéron defended a scientific approach to behaviour, which did not assume any particular 'communality' between the subject and object of research, just as the physicist does not expect nature to behave according to human desires and finality.

In different ways, the social sciences as represented by the Durkheimians and by Lévy-Bruhl challenged traditional history of philosophy. The history of the mind either as continuous evolution or as discontinuous and fragmented reality could not be easily reconciled with the idea of the *philosophia perennis* that allowed historians of philosophy ideally to be contemporaries of Plato or Aristotle. Lévy-Bruhl's theory was in many ways more radical than a theory of evolution, which had more established philosophical precedents in the theories of progress. It opened the doors to relativism and broke the universality of the human mind. These were no minor matters. The possibility of a direct engagement with past texts was the assumption that justified the mainstream manner of approaching the history of philosophy. In the following chapters, we shall see that this type of relativism, much less radical than present-day cognitive relativism, as Warren Schmaus has observed,[48] proved methodologically fecund for historians of science who wanted to avoid anachronism and discarded the model of the history of science as continuous progress. Before examining their historiography, an analysis of Léon Brunschvicg's theories of history and of the mind is in order. The successor of Lévy-Bruhl to the Sorbonne chair of History of Modern Philosophy, he occupied a crucial position in the intellectual development of the approaches to the history of the mind, from its roots in history of philosophy to the historical epistemology of Gaston Bachelard and Georges Canguilhem.[49]

Léon Brunschvicg's History of the Mind

The History of Philosophy and the History of Science as the Philosopher's Laboratory

Léon Brunschvicg's philosophy may at first sight seem as contradictory as his social and academic position. His comparatively humble origins have to be reconciled with the high bourgeois individual of some reports, including Simone Weil's,[50] and with

47 The use of introspection in psychology had been the object of lengthy debates, see: Martin Kusch, *Psychological Knowledge: A Social History and Philosophy* (London and New York: Routledge, 1999).

48 Schmaus, 'Lévy-Bruhl, Durkheim, and the Positivist Roots of the Sociology of Knowledge', p. 457.

49 I agree with Frédéric Worms, who has argued that, in order to understand twentieth-century reflection on science, it is necessary to go back to Bergson and Brunschvicg, for their doctrines are at the origin of the two major views of science: that of Deleuze and Merleau-Ponty, and that of Bachelard and Cavaillès, respectively: see Frédéric Worms, 'Between Critique and Metaphysics: Science in Bergson and Brunschvicg', *Angelaki* 10/2 (2005): 39–57 (39).

50 Quoted in Pétrement, *La vie de Simone Weil*, vol. 1: *1909–1934*, p. 305.

the philosopher who frequented the salon of Madame Ludovic Halévy (Elie's sister-in-law), and who reportedly could be met at modern art exhibitions as well as the Louvre, and at concerts of both classical and contemporary music.[51] Similarly, his philosophy may well puzzle a reader who hopes to draw clear distinctions between philosophers and social scientists, between historians of philosophy and historians of science, and between the advocates and opponents of evolution and indeed progress in intellectual history.

Brunschvicg used Lévy-Bruhl's theories in his own books, and emphasized the latter's links to the history of philosophy, rather than to the social sciences. For Brunschvicg, Lévy-Bruhl was 'free from the dogmas of the [sociological] school' precisely because his roots were in the history of philosophy, as his works on Jacobi and Comte showed.[52] Notably, in Brunschvicg's account, Lévy-Bruhl was free from two sociological dogmas. One is the assumption that the mind and its categories are the same in all times and places; the second is the interpretation of different ways of thinking simply as stages of a universal mental evolution.[53] There is no doubt that Brunschvicg approved of Lévy-Bruhl's positions on these matters. Nevertheless, not only did he write books on the 'stages of mathematical philosophy' and the 'ages of intelligence', and an article on a 'phase of mathematical development',[54] but he also employed Lévy-Bruhl's accounts of the 'prelogical' approach to numeration in order to describe the early concept of number,[55] so interpreting Lévy-Bruhl's primitive mentality as the mentality of human beginnings. At the same time, he claimed to be more indebted to Lévy-Bruhl's work than to that of other ethnologists, precisely because the latter had established an opposition between our logical habits and primitive mentality.[56]

A significant part of Brunschvicg's work is a reflection on the history of science; this was not regarded favourably by a number of his colleagues. At a meeting of the Société française de philosophie dedicated to 'history and philosophy', Dominique Parodi simply stated what certainly many of his colleagues thought, that is, that Brunschvicg, in his 'latest work',[57] appeared to have given up philosophy in favour of a history of scientific concepts.[58] Etienne Gilson was not present, but elsewhere

51 Fabiani, *Les philosophes de la République*, p. 116, quoting Marcel Deschoux, *Léon Brunschvicg ou l'idéalisme à hauteur d'homme* (Paris: Seghers, 1969).

52 Léon Brunschvicg, 'Nouvelles études sur l'anime primitive', *Revue des deux mondes* 52 (1932): 172–202 (174).

53 Ibid.

54 Léon Brunschvicg, *Les âges de l'intelligence* (Paris: Alcan, 1934); Léon Brunschvicg, *Les étapes de la philosophie mathématique* (Paris: Alcan, 1912); Léon Brunschvicg, 'Une phase du développement de la pensée mathématique', *Revue de métaphysique et de morale* 18 (1909): 309–56.

55 Brunschvicg, *Les étapes de la philosophie mathématique*, ch. 1: 'L'ethnographie et les premières opérations numériques'.

56 Ibid., p. 7.

57 Parodi was referring to Léon Brunschvicg, *L'expérience humaine et la causalité physique* (Paris: Alcan, 1922).

58 Parodi in Léon Brunschvicg et al., 'Histoire et Philosophie. Séance du 31 mai 1923', *Bulletin de la Société française de Philosophie* 23 (1923): 145–72 (167–8).

wrote that 'the idealism of Monsieur Brunschvicg … reduces philosophy to a critical consideration of the stages passed through by thought in the constituting of science – the history of Mind'.[59] Brunschvicg's new approach might have surprised some, because his previous production perfectly fitted with what was expected from a historian of philosophy, as exemplified by articles and books on Spinoza[60] and a critical edition of Pascal's *Pensée*,[61] in addition to more generally philosophical works, such as his thesis *La modalité du jugement*, and *Introduction à la vie de l'esprit*.[62]

For most historians of philosophy, the history of philosophy was a philosophical activity precisely because it was different from the history of science. They argued that the latter exhibits progress and therefore old doctrines are irrelevant to modern research, whereas the work of the history of philosophy is part of the *philosophia perennis*; as a consequence, the modern philosopher is able to philosophize together with the old masters. A case in point is the disagreement between Robin and Berr to which the latter hinted during the meeting of the Société française de philosophie discussed in Chapter 2. Berr was referring to the controversy that they had when he invited Robin to write a history of ancient philosophy and science together with Abel Rey, for his series *L'évolution de l'humanité*. However, while Berr and Rey believed that the history of philosophy and science could be written as part of a single intellectual development, Robin strongly disagreed. The result was that Robin wrote *La pensée grecque et les origines de l'esprit scientifique*,[63] and spelt out his position in the Introduction. For him, while the problems posed by Greek philosophers were still alive, and the history of philosophy is 'philosophy itself', the history of science could only satisfy 'the scholar's curiosity' rather than 'thought's most general and profound needs'.[64] Berr explained his opposite point of view in the Preface of Robin's monograph. In the Preface of Rey's volume on ancient oriental science in the same series, Berr also gave details of the disagreement that had occurred between them and announced that Rey would write other volumes, in order to cover the Greek

59 Etienne Gilson, 'Concerning Christian Philosophy: The Distinctiveness of the Philosophic Order', in Raymond Klibansky (ed.), *Philosophy and History* (Oxford: Clarendon Press, 1936), pp. 61–76 (70). Eleven years earlier, Gilson described Brunschvicg as his 'teacher and friend': 'Une heure avec Etienne Gilson' [1925], in Lefèvre, *Une heure avec …*, p. 68.

60 See Chapter 2, n. 89 above.

61 Blaise Pascal, *Pensée et opuscules, publiés avec une introduction, des notices, des notes, par M. Léon Brunschvicg* (Paris: Hachette, 1897).

62 Léon Brunschvicg, *La modalité du jugement* (Paris: Alcan, 1897); Léon Brunschvicg, *Introduction à la vie de l'esprit* (Paris: Alcan, 1900). The earlier work, which was one of his doctoral dissertations, was republished in the 1960s together with the translation from Latin of his other dissertation: Léon Brunschvicg, *La modalité du jugement, troisième édition augmentée de La vertu métaphysique du syllogisme selon Aristote, thèse latine traduite par Yvon Belaval* (Paris: Presses Universitaires de France, 1964).

63 Léon Robin, *La pensée grecque et les origines de l'esprit scientifique* (Paris: La Renaissance du livre, 1923).

64 Ibid., p. 5.

period, so as to remedy what Berr regarded as the limitations of Robin's approach.[65] Rey readily obliged, and another four volumes on ancient science followed.[66]

In this disagreement, the positions were relatively clear, for Abel Rey, although a philosopher, had made a choice in favour of the history of science as a discipline. Brunschvicg, however, never made that choice, although some could have been led to believe that he had. When George Sarton, the founder of *Isis*, invited him to collaborate more closely with his journal, Brunschvicg replied that, despite his research in the field of the history of physics and mathematics, in fact he was not a historian of science, but rather a scholar whose research was in the field of consciousness and philosophy of science.[67] His 'faithfulness' to philosophy did not protect him from the implicit or explicit suspicions of those colleagues who, like Parodi, were more inclined to a clearer separation between the history of philosophy and the history of science. The importance that Brunschvicg accorded to science seemed to many not appropriate for a philosopher. Indeed, Célestin Bouglé suggested that the Catholic philosopher Maurice Blondel was probably referring to Brunschvicg when he condemned the attitude that makes philosophy 'a servant of the sciences'. Unlike Blondel, Bouglé regarded Brunschvicg as the philosopher who had 'broken with the "literary" philosophical tradition' that had dominated French universities since Cousin, and had 'courageously' and competently studied the sciences.[68] Bouglé, Professor of History of Social Economy and founder and Director of the Centre de documentation sociale at the Ecole normale supérieure, where he was also Associated Director,[69] was a social scientist, though one who stayed particularly close to his philosophical roots. Brunschvicg on the other hand firmly stayed within philosophy and was never tempted to cross over to the social sciences.

Brunschvicg's overarching aim was to study the mind, and his method was to study it *a posteriori*, that is to say by observing its activities. For him, the philosophers needed 'a laboratory' in which to observe the mind at work, and this was history; he claimed that history was for the philosopher just what the laboratory was for the scientist.[70] The history of philosophy and the history of science were both excellent

---

65 Henri Berr, 'Avant-propos: La science et la genèse de la raison', in Abel Rey, *La science dans l'antiquité, vol.1: La science orientale avant le grecs* (Paris: La Renaissance du livre, 1930), pp. vi–vii.

66 Abel Rey, *La science dans l'antiquité, vol. 2: La jeunesse de la science grecque* (Paris: La Renaissance du livre, 1933); Abel Rey, *La science dans l'antiquité, vol. 3: La maturité de la pensée scientifique en Grèce* (Paris: La Renaissance du livre, 1939); Abel Rey, *La science dans l'antiquité, vol. 4: L'apogée de la science technique grecque: Les sciences de la nature et de l'homme, les mathématiques d'Hippocrate à Platon* (Paris: La Renaissance du livre, 1946); Abel Rey, *La science dans l'antiquité, vol. 5: L'apogée de la science technique grecque. L'essor de la mathématique* (Paris: La Renaissance du livre, 1948).

67 Brunschvicg, Letter to Sarton, 2 February 1923, Houghton Library, bMS Am 1803 (218).

68 Célestin Bouglé, *Les maîtres de la philosophie universitaire en France* (Paris: Maloine, 1938), p. 63.

69 Bouglé became director of the Ecole normale supérieure in 1935.

70 Brunschvicg et al., 'Histoire et Philosophie. Séance du 31 mai 1923', p. 162.

laboratories as far as he was concerned. In a time when experimental psychologists were developing their laboratories more and more, and there studied the mind, one could have wondered why Brunschvicg did not even consider a real laboratory rather than an ideal one. The answer is implicitly given by his philosophy: the philosophers who studied the mind *a priori* and the psychologists who studied the mind in their laboratory and reached general conclusions, ultimately shared the assumption that the mind does not change. Brunschvicg, with Lévy-Bruhl, did not believe in the fixity of the mind. He regarded his view of the mind as part of his neo-Kantianism; he argued that he was being 'more faithful than Kant to the spirit of critical idealism', by rejecting the table of categories as given once and for all.[71] He did not present the variability of mental categories as an assumption, but rather as the result of his study of the history of thought. Just as Lévy-Bruhl was convinced of the existence of different mentalities by his study of the so-called primitive, Brunschvicg was particularly struck by the revolutions of modern science. For him, non-Euclidean geometry had shaken the conception of space that philosophers assumed as inherent to the human mind, and Einstein's theory of relativity had spared neither the conception of space nor that of time. Brunschvicg argued that the Kantian conditions of our sensible intuition, absolute space and time, had been shown not to be the only possible 'containers' of our experiences.[72] Indeed, as he told Einstein in person at a meeting of the Société française de philosophie, he believed that the theory of relativity had gone beyond the Kantian distinction between the form and the content of knowledge.[73] In other words, for him modern science has shown that there is no longer an abstract space – or the space of Euclidean geometry – in which objects are placed, just as there is no abstract time in which our experiences are placed. In his view, Kant, while claiming to have described the '*intellectus archetypus*' and 'the immutable structure of the mind', was in fact describing the form of Newtonian

71 Brunschvicg, *L'expérience humaine et la causalité physique*, p. 550. The variety of neo-Kantian philosophies and their views of mental categories is great: for a general study of (German) neo-Kantianism, see Klaus Christian Köhnke, *The Rise of Neo-Kantianism* (Cambridge: Cambridge University Press, 1991). On Brunschvicg's neo-Kantianism, see Simone Goyard-Fabre, 'Léon Brunschvicg et l'héritage kantien', in Jean Ferrari, et al. (eds), *Kant et la France – Kant und Frankreich* (Hildesheim, Zurich and New York: Georg Olms, 2005), pp. 261–9. On neo-Kantianism in French philosophy of science, and mathematics in particular, see Hourya Benis Sinaceur, 'From Kant to Hilbert: French Philosophy of Concepts in the Beginning of the Twentieth Century', in José Ferreirós and Jeremy Grey (eds), *The Architecture of Modern Mathematics* (Oxford: Oxford University Press, 2006), pp. 349–76.

72 Brunschvicg, *L'expérience humaine et la causalité physique*, p. 458.

73 Einstein's reply must have disappointed Brunschvicg. Considering himself a Kantian, he said that 'each philosopher has his own Kant', and he did not know which Kant Brunschvicg's was. He also added that he subscribed to Kant's theory because it showed that in science there are *a priori* concepts on which to build science. He believed that the only alternative was to claim either that there are *a priori* concepts in our consciousness as Kant did, or that these are conventional, as Poincaré claimed: Einstein, 'La théorie de la relativité. Séance du 6 avril 1922', pp. 101–2.

mechanics.[74] Brunschvicg went so far as to call the belief in immutable *a prioris* of knowledge a superstition.[75]

His criticism of Kant explains why he needed to analyse the history of thought. If there are no categories given once and for all, and if the ways in which we think change, it is not possible to reach general conclusions about the mind by analysing it only in one particular time and place. For Brunschvicg the philosopher should reflect on the history of science, not simply on science in a particular moment of its development. Indeed, he almost mocked nineteenth-century philosophers of science who struggled to define concepts and to isolate fundamental principles, only for science to discard those concepts and principles. For him, the essence of philosophy is to grasp the mind's activities and changes in history: the history of philosophy then becomes the history of the growth of the philosophical mind.[76]

The Troubled History of Progress

Needless to say, Brunschvicg's conception of the history of philosophy as a 'narrative', or as a history observed in its unfolding in time, could not satisfy his colleagues who located the philosophical character of the history of philosophy precisely in the timelessness of the texts that made it up. The objections that more recently Martial Gueroult has made to Brunschvicg illustrate the point of view of mainstream historians of philosophy, including Brunschvicg's own colleagues. Gueroult has claimed that from Brunschvicg's conception of philosophy as reflection on the history of thought it follows that the history of philosophy is in effect expelled from philosophy. He has argued that for Brunschvicg the object of philosophy is the history of the mind and its evolution; as a consequence, the various doctrines that make up that history are for him only documents of a certain 'mental reality', whereas their explicit content loses its philosophical interest.[77] In Brunschvicg's hands, the history of philosophy is a history in the sense of development and change rather than heritage. For him, philosophy does not enjoy an 'eternal present', following Bréhier's expression, but it is affected by time just as much as science.

There was one point, though, on which Brunschvicg totally agreed with other historians of philosophy, including Delbos, Boutroux, Bréhier and Robin: the history of thought does not follow laws of development. He rejected any philosophy of history that included a prediction of future developments, including Hegel's, Marx's and Nietzsche's. He also rejected:

> Those magic formulae, at once simple and universal, which, following the example of the romantics and the dialecticians, the so-called positivists of the nineteenth century

74 Brunschvicg, *L'expérience humaine et la causalité physique*, p. 552; Brunschvicg et al., 'Histoire et Philosophie. Séance du 31 mai 1923', p. 147.

75 Brunschvicg, 'History and Philosophy', p. 33.

76 See Martial Gueroult et al., 'Commémoration du Xe anniversaire de la mort de Léon Brunschvicg: Brunschvicg et l'Histoire de la Civilisation. Séance du 30 janvier 1954', *Bulletin de la Société française de Philosophie* 48 (1954): 1–36 (6).

77 Martial Gueroult, *Dianoématique, livre II: Philosophie de l'histoire de la philosophie* (Paris: Aubier, 1979), pp. 22–3.

> – biologists or sociologists – have so abundantly elaborated, and which nowadays interest nobody except the authors of school text-books.[78]

Although he referred to nineteenth-century sociology, bearing in mind that he praised Lévy-Bruhl for avoiding the 'dogmas' of the sociological school, it is clear that Brunschvicg regarded sociologists with a certain amount of suspicion, in particular because in his eyes they limited human beings' freedom in creating their future. He concludes his article with the these words:

> The definitive ruin of the systems which were at once pretentious and fragile, far from injuring the cause of rational speculation, serves to introduce us to a philosophy of the human mind in which history, taken by itself, is not only an instrument of knowledge but the very thing which must be grasped in the continuity of its development, in the fertility of the resources which it has prepared for a free working out of the future.[79]

In Brunschvicg's account, there may be no laws of historical development, but certainly there is an empirically observed progress. His interest did not lie only in drawing the lines along which the human mind in his view had developed and changed, but also in judging this development. There is no question that Brunschvicg was not a relativist when it came to evaluating the various moments of intellectual history. He judged some moments in history as progressive, and others as regressive. Socrates and Descartes represented for him important moments of departure in the advancement of European civilization.[80] On the other hand, he regarded Aristotle's philosophy as almost a return to a primitive mentality; he even claimed that it reminded one of the child's mentality as described by the psychologist Jean Piaget.[81] Indeed, Brunschvicg claimed that Aristotle almost compromised all the progress that intelligence had made after Thales.[82] And again, while the Enlightenment was for him a crucial stepping-stone in the progress of the European mind, nineteenth-century German romanticism and idealism were retrograde movements.[83]

In order to judge, it is necessary to have norms on which to base one's judgements. It is not too difficult to work out which norms Brunschvicg used in his evaluation of doctrines, and his reconstruction of the progress and setbacks that human reason experienced in its history.[84] The most important norms include the mathematization of knowledge, idealism, secularism and self-reflectivity. He argued that mathematics

---

78 Brunschvicg, 'History and Philosophy', p. 33.

79 Ibid., pp. 33–4.

80 Brunschvicg, *L'esprit européen*, p. 17; Léon Brunschvicg, 'Introduction: L'humanisme et l'occident' [1928] in *Ecrits philosophiques* (Paris: Presses Universitaires de France, 1951); Brunschvicg, *L'expérience humaine et la causalité physique*, p. 572; Léon Brunschvicg, 'L'orientation du rationalisme: Représentation, concept, judgement', *Revue de métaphysique et de morale* 27 (1920): 261–343 (342).

81 Brunschvicg, *Ecrits philosophiques*, pp. 2–3.

82 Brunschvicg, *L'esprit européen*, p. 47.

83 Ibid., pp. 138–48.

84 It is more difficult to find a justification for his norms. In the next chapter, we shall see how Bachelard and Canguilhem justified the norms that they respectively employed. Brunschvicg did not offer, as far as I can see, a satisfactory justification.

is the only language in which nature agrees to respond to our questions.[85] For him those thinkers who saw the world in mathematical terms, or dedicated themselves to the study of mathematics, positively contributed to the advancement of knowledge: Pythagoras and Descartes imparted a forward movement to the history of thought, whereas Aristotle, by abandoning mathematics, returned to 'common-sense realism', which for Brunschvicg is the mark of the type of thought that Lévy-Bruhl called prelogical and prescientific.[86] Whereas realism is for Brunschvicg a dogmatic and naïve attitude, mathematics is the expression of an idealist approach to knowledge, for it is an activity in which the mind operates in a particularly independent manner. He considered himself 'idealist, totally idealist',[87] and judged as naïve the belief that there is a dichotomy between ideas on the one hand and mind-independent things on the other. If these independent things exist, he argued, we cannot say anything about them.[88] Similarly, the idea that truth is something independent of us, and external to us, is for him a naïve conception. In his view truth has its source in the mind, for it is the mind that establishes intelligible relationships between phenomena.[89] For Brunschvicg neither nature nor the mind is given once and for all: nature is nature as we experience it; and since for him the way in which we organize the data of our experience changes, nature will also change.

Brunschvicg's idealism places the human being at the core of all aspects of his epistemology, and of his philosophy as a whole. Transcendent truth and religious truth play no role in his philosophy. Indeed, another norm by which he judged progress is secularization [*laïcisation*]. It is apparent that secularism is for him the achievement of the 'European mind', which he not only counterpoised to the 'primitive mind' as presented by Lévy-Bruhl, but also to the 'eastern mind'.[90] Unoriginally, he contrasted the ancient 'eastern cosmogonies' based on religious faith to the emerging western philosophy of Ionian thinkers such as Thales and Heraclitus. He argued that the latter, although still intuitive and not scientific, was nonetheless based on reason. Similarly, reason alone, rather than revelation or tradition, guided for him Pythagoreanism insofar as it introduced rigorous demonstrations into arithmetic. However, in his view the first 'real revolution' of the European mind after Socrates had taken place with Descartes and Galileo.[91] For him, it was in the seventeenth century that a truly rational and secular approach to knowledge emerged.

Brunschvicg saw the seventeenth century's new cosmology as the triumph of mathematics and objective experience, which worked independently of any dogma and revealed truths, and ousted superstition. He regarded this new 'speculative reason' as just as autonomous and 'disinterested' as 'practical reason' had been in

85 Brunschvicg, *Ecrits philosophiques*, p. 7. Brunschvicg was writing in 1928.

86 Brunschvicg, *L'esprit européen*, p. 50.

87 Brunschvicg et al., 'Histoire et Philosophie. Séance du 31 mai 1923', p. 170.

88 Léon Brunschvicg, *L'idéalisme contemporain* (Paris: Alcan, 1921), pp. 47–8.

89 Ibid., pp. 86–7.

90 Brunschvicg, *L'esprit européen*, pp. 17–8.

91 Ibid., pp. 17, 82–3. Brunschvicg, 'L'humanisme et l'occident' [1928], in Brunschvicg, *Ecrits philosophiques*.

Socrates' example.[92] For Brunschvicg scientific rationality is a type of humanism precisely because it is independent of any dogmas and transcendent truths. Indeed, for him the theory of relativity was deeply humanistic because it eliminated the single space and single time of classical physics; nobody could say any longer, Brunschvicg argued, that God has given us space and time to contemplate 'in their immense emptiness'.[93] Everything for him is the result of human activity and human conquest, including those very concepts that had seemed to form the structure that makes human knowledge possible, that is to say Kantian *a priori* intuitions of space and time, and categories. It comes as no surprise that Brunschvicg denied the possibility of Christian philosophy; at Gilson's seminar on the notion of Christian philosophy, he argued that Christianity implies a transcendent truth, and transcendent truth is the negation of philosophical activity.[94] Brunschvicg's views on God and religion sparked a disagreement ('*la querelle de l'athéisme*') between himself on the one hand, and Gilson and other Christian philosophers on the other.[95] Brunschvicg seemed not to want to leave God to the followers of positive religions: he vindicated for himself the 'God of philosophers', as opposed to a transcendent and personal God. The progress of science and the retreat of the God of theologians for Brunschvicg went hand-in-hand: Galilean cosmology did not just propose a new astronomical model, but it overcame the 'naively egocentric' vision of the world on which old metaphysics and theologies were based. He presented positive religions and tradition as the main threats to western civilisation and its progress. A few examples will give an idea: although he praised the role of Pythagoras' teaching, his legacy was for him ambiguous in the split among his followers between 'learners', or mathematicians, and 'hearers' who aimed to preserve their master's secret teaching.[96] Similarly, Plato's 'rational idealism' was for Brunschvicg a great triumph of western thought, indeed one of his highest moments,[97] but Plato also reintroduced myths

92 Léon Brunschvicg, 'Religion et philosophie', *Revue de métaphysique et de morale* 42 (1935): 1–13 (9).

93 Léon Brunschvicg, 'Du XIXe au XXe siècle, Troisième partie: L'humanisme de l'intelligence', *Revue de Paris* (February 1927): 576–91 (586).

94 Brunschvicg in Etienne Gilson et al., 'La notion de philosophie chrétienne. Séance du 21 mars 1931', *Bulletin de la Société française de Philosophie* 31 (1931): 37–93 (73). For this debate about Christian philosophy, see also Emile Bréhier, 'Y a-t-il une philosophie chrétienne?', *Revue de métaphysique et de morale* 38 (1931): 133–62; Gilson, 'Concerning Christian Philosophy'; Maurice Blondel, 'Y a-t-il une philosophie chrétienne?', *Revue de métaphysique et de morale* 38 (1931): 599–606.

95 See Léon Brunschvicg, 'Mathématique et métaphysique chez Descartes', *Revue de métaphysique et de morale* 34 (1927): 277–324; Léon Brunschvicg et al., 'La querelle de l'athéisme. Séance du 24 mars 1928', *Bulletin de la Société française de philosophie* 28 (1928): 50–95; Gilson et al., 'La notion de philosophie chrétienne. Séance du 21 mars 1931'; Brunschvicg, 'Religion et philosophie'. Brunschvicg also attracted lengthy attacks published in *Archives de philosophie*: André Bremond, 'Quelques essais de "religion" rationaliste: Alain, J. Bois, L. Brunschvicg, E. Gilson, D. Parodi, Th. Ruyssen', *Archives de philosophie* 8/4 (1931): 60–117; André Bremond, 'Rationalisme et religion', *Archives de philosophie*, 11/4 (1935): 1–203.

96 Brunschvicg, *L'esprit européen*, p. 21.

97 Ibid., p. 53.

in his *Republic*.[98] Even the philosopher to whom Brunschvicg arguably felt closer, Baruch Spinoza, was not completely clear on religious values. On the one hand, he applied 'positive' methods to the exegesis of the Bible in his *Tractatus theologico-politicus*, and discarded the 'vulgar beliefs' of cosmic finality and immortality; on the other hand, in the fifth book of his *Ethics*, he reintroduced religious values.[99]

As well as secularism, Brunschvicg also employed the norm of self-reflectivity in order to evaluate the progress of intellectual and moral history. Since he denied the existence of transcendent norms and values, philosophical, scientific and moral progress is possible for him only by reflecting on the mind itself. However, in his account self-reflectivity is very far from any type of subjective introspection, and it is always rational. In his account, self-knowledge is never knowledge of one's subjective individuality, but rather knowledge of the human mind in its concrete evolution through the centuries. For him, the champion of self-reflectivity was Socrates.[100]

In the last lectures that he delivered at the Sorbonne (1939–40) before retiring, and shortly before going into hiding, where he died in 1944, he gave his uncompromising support to western thought, which for him never stands still.[101] The progress that he described not only in his last lectures, but in much of his work on the history of philosophy, mathematics and physics, is never continuous or assured. The way in which civilizations progressed or failed to do so was not only an abstract historiographical issue. Brunschvicg had to reconcile his support of western thought and his belief in its progress with Nazism and the war that Germany was waging on other countries. He had to reconcile the Germany of Leibniz, Goethe, Schiller and Kant with the Germany that was threatening European civilization. Indeed, he presented his lectures as a repository of the riches that Great Britain and France had been called to defend.[102] Despite the dramatic setting of his lectures, their content does not differ from what he had already argued in his previous works, and it would be wrong to read this book only in relation to the contemporary international events. Contemporary events, however, found a place in his overall view of history, as he regarded them as an example of the recurrent decadence of reason. He drew a parallel between what had happened in Germany and what had previously taken place in France. In France, he argued, the Enlightenment had offered the most admirable achievement of human reason: Condorcet's work, despite being seen by many as utopian, was for Brunschvicg a work of positive science; he admired in particular the use that Condorcet intended to make of mathematics, as a type of inquiry that could overcome intellectual inequalities, and contribute to justice. But of course, he pointed out, the Enlightenment and the French revolution had been followed by Bonaparte's dictatorship.

---

98 Ibid., p. 142.

99 Ibid., p. 100; Brunschvicg, 'Sommes-nous spinozistes?' [1927], in *Ecrits philosophiques*, pp. 156–7.

100 Brunschvicg, *L'esprit européen*, pp. 52–3. For Brunschvicg's view of self-knowledge, see Léon Brunschvicg, *De la connaissance de soi* (Paris: Alcan, 1931).

101 Brunschvicg, *L'esprit européen*, p. 183.

102 Ibid., p. 7.

In Germany, he continued, Leibniz first and then Kant had made European civilization progress, but this advancement had been followed by a decadence 'of intelligence and morality' that eventually opened the way to the 'brutality of animal instinct'.[103] Brunschvicg thought that such decadence had had its first beginnings in Kant's philosophy itself, when, in the 'transcendental dialectic' he reintroduced psychology, cosmology and theology that he had previously excluded from the domain of knowledge.[104] Once again, for Brunschvicg what escapes rational analysis is dogmatic, and for him Kant's philosophy, which had started as revolutionary, ended up being a restoration of old dogmas. He argued that Romanticism, along with Fichte's and Hegel's philosophies, were further stages of a decadence whose dramatic results were under everybody's eyes. In his view, the war that had started at the core of Europe was one dramatic example of the battle between rationality and irrationality, and between progress and reaction.

Social and Disciplinary Mobility

Between 1908 and 1940 the Sorbonne chair of History of Modern Philosophy was occupied first by Lucien Lévy-Bruhl and then by Léon Brunschvicg. However, it was not the case that the latter started teaching at the University of Paris when the other had left: Brunschvicg became *maître de conférences* at the Sorbonne in 1909, and Lucien Lévy-Bruhl continued his activities after his retirement in 1926, not only as professor emeritus, but as director of the Institut d'Ethnologie. The continuity between them is reinforced by the fact that, as I have mentioned, the former seemed to have played an important role in the latter's appointment to a chair. These two professors appear to differ from their colleagues in many ways, notably in their social background: their families of origin were the least wealthy among the Parisian historians of philosophy appointed at the Sorbonne in the first half of the twentieth century, as I have discussed in Chapter 1. To their socially modest origins their Jewish extraction must of course be added. Although both of them became very important academics in the most important French university, their origins in many ways set them apart from 'mainstream' society, and certainly from the Catholic bourgeoisie. Their success could not be linked to a social 'reproduction', as they were breaking new ground as far as their social origin was concerned. Their social networks were different from those of the established intellectual classes; their success apparently due to their own merit and to the French educational institutions.

Because of their social origin, it is a safe guess to say that they would not have had any chance of success had they not being admitted to the most prestigious educational establishments: prestigious Parisian *lycées* and the Ecole normale supérieure. These schools were not only sites of learning, but also of networking, and both of them made friends there who influenced their future in very different ways. Brunschvicg, as discussed above, met Xavier Léon and Elie Halévy at the *lycée* Condorcet, and became part of the group that planned and realized the *Revue de métaphysique et*

103 Ibid., p. 142.

104 Ibid.

morale, the Société française de philosophie and the International congresses of philosophy. His school friends were Jewish like himself, but they had the great advantage of economic capital. Lévy-Bruhl met Jean Jaurès, the future member of parliament and leader of the Socialist Party, at the Ecole normale supérieure.[105] Their friendship did not mark Lévy-Bruhl's academic activities, but rather the elaboration of his political beliefs.[106] This socialist Sorbonne professor, apparently respected and liked by allies as well as opponents, for all his life was 'devoted' to Jaurès and to the proletariat, as Marcel Mauss put it.[107] Although he never formally joined the Socialist Party, not only did he support its causes ideally, but also financially; notably, he generously funded *L'Humanité*, which started as the newspaper of the socialists under the Jaurès' editorship (1904–1914).[108]

Brunschvicg's politics were less on the left than Lévy-Bruhl's, and his engagement less vigorous, but his outlook was certainly progressive. As mentioned in Chapter 1, his wife Cécile, née Kahn, was the president of the Union française pour le suffrage des femmes and a member of the Radical Party,[109] and was appointed Education junior minister in the Blum government. Léon Brunschvicg was a member of the management committee of the League des droit des hommes, was a promoter of women's rights and we find his name among the signatures to various petitions, an important indicator in French politics. For instance, when *L'Humanité* launched a campaign, which it presented as 'the new Dreyfus case', to save the trade unionist Durand who had been condemn to death, Brunschvicg supported it with his signature, so joining, among others, Lévy-Bruhl, Gaston Milhaud, Paul Fauconnet, Victor Basch, Lucien Febvre and Alfred Dreyfus himself.[110] When, in 1913, *L'Humanité* published an appeal against the law that extended the military service to three years, Brunschvicg's signature was in the first group, along with those of other colleagues

---

105 About the Ecole normale supérieure and politics, see Sirinelli, *Génération intellectuelle*.

106 Lévy-Bruhl also wrote a biography of Jaurès: Lucien Lévy-Bruhl, *Jean Jaurès. Essai biographique. Nouvelle édition suivie de lettres inédites* (Paris: Rieder, 1924).

107 Quoted in Marcel Fournier, *Marcel Mauss* (trans. Jane Marie Todd; Princeton and Oxford: Princeton University Press, 2006 [1994]), p. 122.

108 *L'Humanité* later became the newspaper of the French Communist Party, when this was created following the split in the Socialist Party. The paper's title varied accordingly: from *L'Humanité: journal socialiste quotidien* (1904–1919) to *L'Humanité: journal communiste* (1920–23), and, in 1923, to *L'Humanité: organe central du Parti communiste français*.

109 It is quite difficult to define precisely the Radical Party, as the differences between its internal left and right could be quite considerable, and at times it was closer to the left, while at others to the right. With some approximation, one can say that it was stronger among the petite bourgeoisie, and its members supported progress and equal opportunities, but, unlike the socialists, did not seek a radical transformation of society, see R. Vandenbussche, 'Parti Radical', in Jean-François Sirinelli (ed.), *Dictionnaire historique de la vie politique française au XXe siècle* (Paris: Presses Universitaires de France, 2003), pp. 895–906.

110 For the text of the petition, including Milhaud's, Fauconnet's, Basch's and Herr's signatures, see *L'Humanité* (1 January 1911), p. 1; the title 'La nouvelle affaire Dreyfus' is in the 4 January issue, Lévy-Bruhl's signature 5 January, Dreyfus's 8 January, Febvre's 11 January and Brunschvicg's 22 January.

at the Sorbonne like Durkheim, Mauss, Bouglé, and also Herr, Parodi and Alain; others, including Lévy-Bruhl's, signed in the following days.[111]

The image of the Third Republic professor (and of the *normaliens* in particular) as bringing forward secular and republican ideals appeared confirmed by the activities of Lévy-Bruhl, Brunschvicg and their colleagues. However, although this image of the Third Republic professor is correct in a very general way, it cannot be applied to all groups. If we examine the Sorbonne and Collège de France historians of philosophy,[112] we find a rather different picture. Albert Rivaud, right-wing Catholic, close to the Action française and briefly minister in the Pétain government, was an exception. However, others among their colleagues, including Laporte, Delbos and of course Gilson at the Collège de France, were practising Catholics. The *querelle de l'athéisme*, spurred by Brunschvicg's thesis on religion, would not make sense if secularism were a general position, and if Christian philosophers, first of all the historian of philosophy Gilson, did not play an important intellectual role in their discipline and beyond. Robin's concerns seemed to have been concentrated on his work, and he does not appear to have expressed political opinions. Rodier and Milhaud, both Dreyfusard, the former from a Catholic family, the latter from a Jewish one, seemed to have more in common with Lévy-Bruhl and Brunschvicg. Like Brunschvicg, Rodier was a member of the Ligue des droits de l'homme (in Bordeaux), and Milhaud signed with the two historians of modern philosophy the petition mentioned above. However they died early in the century, Rodier at the end of the First World War, and the latter just before it.

Although Lévy-Bruhl and Brunschvicg shared with some of their colleagues initiatives, as shown by the two examples above of their petition-signing, and of course academic projects, they appear not to be mainstream, when put in the context of the historians of philosophy. Lévy-Bruhl and Brunschvicg, in different ways, undermined the assumptions on which traditional history of philosophy was based, namely that philosophical questions and ideas are timeless. Indeed, they argued that in different times or societies, there were differences not only in questions and ideas, but indeed in the mental frameworks that make philosophical questions possible. Despite their high competence in exegesis of texts and knowledge of the history of philosophy, they regarded past doctrines as relative to their particular social and historical circumstances. Neither tradition nor religion afforded timeless truths for them. One could say that their own social mobility symbolized the change that they thought had taken place throughout history and that they hoped would take place in the future. They both defended modernity and rationality and regarded them as incompatible with religion and tradition as a value. The common opposition of the *Revue philosophique* and *Revue de métaphysique et de morale* might lead some to think that Lévy-Bruhl and Brunschvicg were on opposite intellectual, political and philosophical camps. As I have shown, this would be untrue. However, differences between them existed and some of them were significant, and not only in their

111 *L'Humanité*, 13 March 1913, p. 1; Lévy-Bruhl's signature appeared three days later. For a discussion of this petition see Sirinelli, *Intellectuels et passions françaises*, pp. 43–5.

112 I am referring to the group analysed in Chapter 1.

conception of history and the mind that I have discussed above, but also in their disciplinary choices.

Pierre Bourdieu, citing Joseph Ben-David and Randall Collins, noted that intellectuals like Durkheim took the risk of investing a large technical capital, which they had acquired in a 'superior' discipline, in a discipline that at the time was considered 'inferior', but they nevertheless manage in time to regain a position equivalent to that associated with the original discipline.[113] Did Lévy-Bruhl and Brunschvicg follow a similar pattern? Their original discipline was certainly very prestigious. Lévy-Bruhl made a choice to invest his knowledge in a new 'inferior' discipline, ethnology, and to use his prestige to establish it at academic level. Although Lévy-Bruhl saw his 'ethnological' studies as a way of answering philosophical questions, they were still perceived as close to sociology, although highly relevant to philosophers, who read and discuss them. He strongly supported the creation of ethnology as an academic discipline, although the foundation of the Institut d'Ethnologie took place rather towards the end of his career. Lévy-Bruhl seemed to have used the prestige attached to himself as a historian of philosophy in order to help the development of the new discipline.

The case of Brunschvicg is rather different. Despite his rather unorthodox relativistic view of the history of philosophy, his alliance was always with philosophy. If Lévy-Bruhl saw continuity between his work in the history of philosophy and his ethnological work, Brunschvicg saw no change at all in his production. He claimed that he investigated different domains of knowledge, including the history of philosophy, mathematics and physics, in order to investigate the mind. He apparently had no interest in founding a new discipline and, as mentioned above, denied being a historian of science, despite being perceived as one. However, similarly to his predecessor as Sorbonne Professor of History of Modern Philosophy, his legacy is not be found prevalently in the history of philosophy. Lévy-Bruhl's conviction that ideas are linked to the culture and society in which they emerge, and Brunschvicg's view that the mind changes throughout time, and that there exists an intellectual progress did not fit with a traditional understanding of the history of philosophy. Other scholars, however, were happy to take up their ideas and apply them to other domains. For all his protestation that he was not a historian of science, Brunschvicg's legacy to the history and philosophy of science was going to be very great. His student Gaston Bachelard, who developed many of his central ideas, did not become a professor of history of philosophy, but rather of history and philosophy of the sciences, and Director of the Institut d'histoire des sciences et techniques, as we shall see in Chapter 6. Lévy-Bruhl's legacy was so widespread that it would be difficult even to mention all the historians, psychologists, ethnologists and other scholars who applied his conception of mentality. His legacy to the history of science was very remarkable, as I shall show in the following chapters.

---

113 Pierre Bourdieu, 'The Peculiar History of Scientific Reason', *Sociological Forum* 6/1 (1991): 3–26 (19). See also J. Ben-David, 'Roles and Innovation in Medicine', *American Journal of Sociology* 65 (1960): 557–68; J. Ben-David and Randall Collins, 'Social Factors in the Origins of a New Science: The Case of Psychology', *American Sociological Review* 31 (1966): 451–65.

Chapter 4

Approaches to the History of the Mind: The History of Science between Philosophy and History

The Diffuse Presence of the History of Science

The academic and intellectual space that the history of science occupied in France in the first half of the twentieth century is not comparable with that occupied by philosophy, nor even with that occupied by the history of philosophy. Although the creation of the first chair of General History of Science at the Collège de France in 1892 benefited from the support of Auguste Comte, this chair had a very tormented history,[1] and the discipline played a relatively marginal role within academia, especially compared with philosophy.[2] The type of history of science that was at the core of the project of writing the history of the mind developed mainly in milieus dominated by philosophers and historians, who often regarded their work partly in

1 For the history of this chair see H.W. Paul, 'Scholarship and Ideology: The Chair of the General History of Science at the Collège de France, 1892–1913', *Isis* 67 (1976): 376–97; Annie Petit, 'L'héritage du positivisme dans la création de la chaire d'histoire générale des sciences au Collège de France', *Revue d'histoire des sciences* 48 (1995): 521–56; and Annie Petit, 'L'enseignement positiviste: Auxiliaire ou obstacle pour l'histoire des sciences?', *Revue d'histoire des sciences* 58 (2005): 329–65. The chair of General History of the Sciences (*Histoire générale des sciences*) was held by Pierre Laffitte between 1892 and 1903, and by Grégoire Wyrouboff between 1903 and 1913. In the inter-war period, the son of Emile Boutroux, Pierre, a mathematician whose project I shall compare with Rey's in this chapter, held the Collège de France chair of History of the Sciences (1920–22). Note the change of name from 'general history of the sciences' to simply 'history of the sciences', which reflects, as we shall see, the convictions of Pierre Boutroux. For a full list of the chairs at the Collège de France since 1530, see the Collège de France website, 'Liste des professeurs depuis la fondation du Collège de France en 1530' http://www.college-de-france.fr/media/lis_prf/UPL45507_LISTE_DES_PROFESSEURS.pdf (accessed 14 November 2007).

2 It hardly needs repeating that Duhem was professor of physics at the University of Bordeaux, and indeed when asked whether he would be available for the chair of History of Science at the Collège de France, he reportedly replied: 'I am a physicist. Paris will obtain me only as such, if I ever should return there' (quoted in S.T. Jaki, *Uneasy Genius: The Life and the Work of Pierre Duhem* [Dordrecht: Martinus Nijhoff, 1987], p. 81). See also Pietro Redondi, *Epistemologia e storia della scienza. Le svolte teoretiche da Duhem a Bachelard* (Milano: Feltrinelli, 1978), p. 27; Hélène Pierre-Duhem, *Pierre Duhem, un savant français* (Paris: Plon, 1936), p. 150; Pierre Humbert, *Pierre Duhem* (Paris: Bluod et Gay, 1932), pp. 18ff.

opposition to that of scientists turned historians and philosophers. In the Faculty of Letters, until the 1930s, the history of science had a very limited space, the only chair related to it being Gaston Mihaud's in History of Philosophy in its Relationship with the Sciences (1909–18), with Abel Rey being *maître de conférences* in Philosophy in its Relationship with the Sciences.[3] Rey, who became professor without chair in 1925, and then Professor of History and Philosophy of the Sciences in 1930, was instrumental in expanding the space of history of science at the Sorbonne by founding the Institut d'histoire des sciences (1932), soon to be renamed Institut d'histoire des sciences et techniques, and its journal *Thalès*.[4] This academic space for the history of science was very important in the development of the philosophical history of science, and of the project of the study of the mind through the history of science. After Rey's death in 1940, his chair and the directorship of the Institut d'histoire des sciences et techniques went to Gaston Bachelard, a student of Brunschvicg and Rey, and then to Bachelard's student Canguilhem.

However, in order to have an idea of the projects that combined the history of science and the study of the mind in the inter-war period, it would not be sufficient to focus on the University of Paris, and indeed academia in general, for at least two reasons. First, this way of approaching science, although successful in terms of creating a tradition, was not at first securely rooted within academia and had a very diffuse presence in different institutions, inside and outside academia. Its practitioners had diverse backgrounds, and various degrees of marginality in terms of gender, nationality and religious origins, which often prevented them from securing academic posts, especially at the Sorbonne. The second reason, which is crucial, is that extra-academic spaces provided an 'injection' of history into the rather philosophical approach to the history of science practised by the Sorbonne philosophers. Rey, who was an academic philosopher, discussed and developed his historiographical ideas within the Centre international de synthèse, an extra-academic institution that turned out to be very important for the history of science. At the Centre Rey closely collaborated not only with its director Henri Berr, but also with the historian of mentalities Lucien Febvre. The Centre international de synthèse also had a Unit for the History of the Sciences (Section d'histoire des sciences), whose 'secretary in perpetuity' (i.e., director) was Aldo Mieli, and its administrator, treasurer and librarian Hélène Metzger. As the latter recorded, the members of Section pledged to promote the history of science in the training of secondary school teachers and to encourage the creation of university institutes devoted to the history of the sciences.[5] It is this second pledge that Rey was able to realize at the University of Paris.

3 See Guigue, *La Faculté des Lettres de l'Université de Paris*, pp. 14–29 (list of chairs), pp. 32–45 (*maîtres de conférences*).

4 Anastasios Brenner has reconstructed the continuity between Milhaud and Rey's teaching and visions of history of science, and their links with Tannery before them: see Anastasios Brenner, 'Réconcilier les sciences et les lettres: Le rôle de l'histoire des sciences selon Paul Tannery, Gaston Milhaud et Abel Rey', *Revue d'histoire des sciences* 58 (2005): 433–54.

5 Hélène Metzger, 'Activité de la Section d'histoire des sciences en 1930–31', *Revue de synthèse* 51 (1931): 192–3 (193).

The Section d'histoire des sciences played a very important role in the development of the history of science in the inter-war period, despite being managed by intellectuals with little power over academic institutions. Important discussions took place there, and, in the following chapter, I shall analyse in particular Hélène Metzger's talks, which not only present her historiography, but also give us an idea of the internal tensions between different ways of conceiving the history of science. We shall see Metzger's approach clash with those of Berr and Mieli. These were not just internal disagreements, in fact they reflected more general ways of conceiving history writing. On the one hand, there were a cluster of projects which, despite their differences, shared a view of the history of science as focused on a comprehensive examination of documents and on the compilation of bibliographies. Not only Berr and Mieli adhered to this view, but also George Sarton, who corresponded with both Mieli and Metzger, and participated in their activities, although from the other side of the Atlantic. Other scholars, notably Metzger and Koyré, aimed to understand ways of thinking and to provide a philosophical interpretation of texts. However, this was not a clear distinction. Lucien Febvre, the Centre co-director, also focused on general ways of thinking, but nevertheless regarded Metzger's approach with a certain amount of suspicion, because in his eyes it undermined historical objectivity.

The Section d'histoire des sciences was not only a place where historians of science such as Metzger and Koyré presented their work. It was also the headquarters of the journal *Archeion*, directed by Mieli, and of the Comité international d'histoire des sciences (also called Académie d'histoire des sciences),[6] an international organization founded and managed by Aldo Mieli, which ran conferences and kept historians of science in touch across the globe.

The Centre international de synthèse as a whole was a formidable place where intellectuals met and collaborated. Its politics were also quite clearly on the left, and it enjoyed the support of distinguished politicians. Henri Berr had been engaged in cultural politics from well before founding his Centre, indeed the latter was one of several media, including his *Revue de synthèse historique*, with which he defended the values of secularism and progress.

The Centre international de synthèse

The Centre international de synthèse (or simply 'Centre de synthèse') was founded by Henri Berr in 1925. It was one of Berr's several important projects, which were linked to one another. He had first created the *Revue de synthèse historique* in 1900 and directed the series of monographs, *L'évolution de l'humanité: synthèse collective*, which came to include such famous books as Lucien Febvre's *Le problème de l'incroyance au XVIe siècle: la religion de Rabelais* and Marc Bloch's

6 It was Hélène Metzger who proposed the name of 'Académie' at an official Comité meeting in 1932; in order to avoid a change in the constitution, she proposed to add the new name as a subtitle to the old one for the time being; her proposal was accepted unanimously: Comité International d'Histoire des Sciences, 'Séance privée du Comité (14 mai, après midi) à la Sorbonne', *Archeion* 14 (1932): 468–73 (472). For the sake of clarity, I shall call this organization the Comité international d'histoire des sciences.

La société féodale.[7] Similarly to the founder of the *Revue de metaphysique et de morale*, Xavier Léon, the founder of the *Revue de synthèse historique* came from the Jewish bourgeoisie, although his family was less affluent than Léon's. Moreover, while Léon was a Parisian, Berr was from Lorraine, and his mother was Alsatian.[8] He attended a Parisian lycée (the Charlemagne, like Léon Robin)[9], the Ecole normale supérieure (admission 1881), obtain his *agrégation* at 21 years of age, and his doctorate in 1899. However, he never held a university post. Despite one of his two candidatures (in 1903 and 1906) to the Collège de France being supported by Henri Bergson, he failed to be elected.[10] He was a *lycée* teacher until his retirement in 1925; notably, he spent 29 years at the prestigious *lycée* Henri-IV, where, at different times, he was a colleague of Brunschvicg, Delbos and Alain. Like Lévy-Bruhl and Brunschvicg, he married a wealthy woman, Cécile Halphen.[11] His marriage conferred on him a higher social status, and a prestigious apartment near the Champs-Elysées where Madame Berr held a literary salon.[12]

---

7 Lucien Febvre, *Le problème de l'incroyance au XVIe siècle: la religion de Rabelais* (Paris: Albin Michel, 1942); Marc Bloch, *La société féodale: la formation des liens de dépendance* (Paris: Albin Michel, 1939); Marc Bloch, *La société féodale: les classes et le gouvernement des hommes* (Paris: Albin Michel, 1940). Febvre co-wrote two other books in the series: Lucien Febvre and Lionel Bataillon, *La terre et l'évolution de l'humanité: introduction géographique à l'histoire* (Paris: La Renaissance du livre, 1922); Lucien Febvre and Henri-Jean Martin, *L'apparition du livre* (Paris: Albin Michel, 1958). Apparently, there were plans for Febvre to write another five or six books in the series, and it was he who decided to publish Bloch's volumes: see Bertrand Müller, 'Lucien Febvre et Henri Berr: de la synthèse à l'histoire-problème', in Agnès Biard, Dominique Bourel and Eric Brian (eds), *Henri Berr et la culture du XXe siècle: histoire, science et philosophie* (Paris: Albin Michel/ Centre international de synthèse, 1997), pp. 39–59 (41).

8 Henri Berr, 'Henri Berr par lui-même', *Revue de synthèse* 85 (1964 [1942]): 1–4 (2). In this brief self-portrait written at the age of 79 years, he indicated his father's profession as that of a 'manufacturer' classified by commentators either as a small industrialist (in Gilles Candar and Jacqueline Pluet-Despatin, 'Introduction: une amitié au service de l'histoire', in Gilles Candar and Jacqueline Pluet-Despatin [eds], *Lucien Febvre: Lettres à Henri Berr* [Paris: Fayard, 1997], p. xi) or as an artisan (in Dominique Bourel, 'Présentation: Henri Berr [1863–1954]', in Biard, Bourel and Brian [eds], *Henri Berr et la culture du XXe siècle*). Berr especially emphasized the activities of his maternal grandfather who founded a newspaper called *La pure vérité*. He said that he had inherited from his grandfather the 'faith in science' and the love for his country: Berr, 'Henri Berr par lui-même', pp. 2–3.

9 Robin was three years younger than Berr. The two appeared to be on friendly terms: Pierre-Maxime Schuhl reported that he was introduced to Berr at Robin's home: Pierre-Maxime Schuhl, '[Hommage à Henri Berr]', *Revue de synthèse* 85 (1964): 135–6 (135).

10 Giuliana Gemelli, 'Communauté intellectuelle et stratégies institutionnelles. Henri Berr et la fondation du Centre International de Synthèse', *Revue de synthèse*, IV series, 2 (1987): 225–59 (231).

11 Cécile Berr's brother Jacques married Marie Durkheim, daughter of Emile.

12 Candar and Pluet-Despatin, 'Introduction: une amitié au service de l'histoire', pp. xii–xiii. Tolédano, general secretary of the Centre de Synthèse (see Bourel, 'Présentation: Henri Berr [1863–1954]', p. 15) wrote that 'without Madame Berr, Henri Berr would not have really been Henri Berr': André D. Tolédano, 'Henri Berr, Le maître et l'ami au jour le jour', *Revue de synthèse* 85 (1964): 144–47 (147).

The Centre de synthèse enjoyed a very large and distinguished membership. Many of the intellectuals and academics mentioned in this book were members of it, although with very different degrees of involvement. These included Gaston Bachelard, Léon Brunschvicg, Jean Baruzi, Emile Bréhier, Alexandre Koyré, André Lalande, Jean Laporte, Lucien Lévy-Bruhl, Marcel Mauss, Hélène Metzger, Emile Meyerson, Dominique Parodi, Henri Piéron and Léon Robin. One of those most involved with Berr's activities was the historian Lucien Febvre, who in 1926 became co-director of the Centre de synthèse.[13] Berr and Febvre also directed the oldest unit of the Centre de synthèse, the Section de synthèse historique (founded in 1926), while the Section des sciences de la nature was first entrusted to another close associate, Abel Rey, and then to the physicist Paul Langevin. In 1931, when the Section de synthèse générale started its activities, Rey obtained its directorship.[14] In the same year, Febvre, Rey and Langevin also joined Berr as editors of the *Revue de Synthèse*, which had become the official publication of the Centre. The journal's wider role was marked by the disappearance of the adjective 'historique' from its title.

Politically, it is not difficult to place the Centre de synthèse; indeed Candar and Plue-Despatin describe it as a 'child of the *Cartel de gauche*'. The very existence of the Centre de synthèse was owed to the support of notable politicians of the *Cartel de gauche*, first of all to the radical member of Parliament Paul Doumer, who was elected speaker of the Senate in 1927 and President of the Republic in 1931. In 1929 he officially opened the new home of the Centre de synthèse, the Hôtel de Nevers (situated opposite the Bibliothèque nationale)[15] which had been offered by the Ministry of Education thanks to his mediation.[16] Doumer remained the chairman of the Centre until 1932, when he was assassinated. As far as political activism goes, there were differences between the members of the inner circle of the *Revue de Synthèse*. In particular, Berr and Rey did not show the same level of personal commitment in left-wing activism as Febvre and especially Langevin did.[17]

13 Candar and Pluet-Despatin, 'Introduction: une amitié au service de l'histoire'. Between 1905 and 1937, Lucien Febvre published some 280 articles and especially reviews in the *Revue de synthèse (historique)*: Müller, 'Lucien Febvre et Henri Berr: de la synthèse à l'histoire-problème', p. 40.

14 See La Rédaction, 'La vie du Centre. La section de synthèse générale. Son programme', *Revue de synthèse* 51 (1931): 127–31.

15 See his speech, along with that of the Education Secretary Pierre Marraud among others, in Centre International de Synthèse, *L'Hôtel de Nevers et le Centre International de Synthèse* (Paris: La Renaissance du livre, 1929).

16 Candar and Pluet-Despatin, 'Introduction: une amitié au service de l'histoire', p. xxx.

17 Febvre, a socialist, was anti-Marxist, unlike Langevin. Before the Second World War, Langevin had a close, though complicated, relationship with the Communist Party, which he only joined in 1944: Bernadette Bensaude-Vincent, *Langevin: science et vigilance* (Paris: Belin, 1987), ch. 13. Langevin was also the promoter, along with Paul Rivet and Alain, of the Comité de Vigilance des Intellectuels Antifascistes (CVIA), of which he was the vice-chairman. Febvre, Lévy-Bruhl and Mauss, among others, joined the Comité: see Sirinelli, *Intellectuels et passions françaises*, p. 144; Bensaude-Vincent, *Langevin: science et vigilance*, pp. 184–92.

Henri Berr was nonetheless very active in cultural politics; for instance he regarded it as his journal's role to support the Sorbonne against attacks from conservatives and indeed reactionaries. The creation of new disciplines, and of the new trends within traditional disciplines, notably French and philosophy, were the targets of conservatives and of those who wanted to defend the French 'literary' tradition. Many articles appeared in the general press, and in papers of the extreme right, such as *L'action française*.[18] The attacks were directed at the 'scientistic' approach of the 'new Sorbonne', and also personally at scholars discussed here, many of whom were members of the Centre de synthèse. In one of a series of articles published in 1910 in *L'opinion*, a certain 'Agathon' (the pseudonym of the student Henri Massis and the young lawyer Alfred de Tarde, as it later emerged) ridiculed the Sorbonne philosophers for their scientific leanings. They argued that philosophy was in the hands, among others, of Durkheim, who was a sociologist and taught a 'social catechism', Lévy-Bruhl, who 'studied savages', Dumas, who 'studied the insane', Delbos, who was an 'excellent historian' (rather than a philosopher), Rodier, who was a philologist, Bouglé, who was 'Durkheim's lieutenant', and Lalande, who was concerned with scientific method.[19] These articles, which were published as a book the following year, had great resonance; they even prompted a discussion in parliament.[20]

Berr contributed his voice, and in marking the tenth anniversary of the *Revue de synthèse historique*, rebuked those attacks, which for him united people with different agendas: some regretted a perceived decadence of literary culture, but others in his view could not 'forgive' the Sorbonne for encouraging democracy and secularism [*esprit laïque*].[21] Indeed, it was not only a matter of conceptions of academic disciplines, but also of politics and indeed anti-Semitism. In the following years, 'Agathon' extended their attacks to the Ecole normale, with the publication of a study the result of which showed, they claimed, that the students were strongly nationalists and they had little time for cosmopolitanism.[22] Although Massis and de Tarde published this report before the First World War, the inter-war period was certainly not free from such attacks. The Action française (to which incidentally Massis

18 Berr provided a list of articles which appeared in 1909 and 1910: Henri Berr, 'Au bout de dix ans', *Revue de synthèse historique* 21 (1910): 1–13 (7, n. 2). Students had started complaining about the new trends at the Sorbonne already at the end of the nineteenth century; for a concise history of these controversies, see Phyllis H. Stock, 'Students versus the University in Pre-World War Paris', *French Historical Studies* 7/1 (1971): 93–110; for a detailed account, see Claire-Françoise Bompaire-Evesque, *Un débat sur l'université au temps de la Troisième République. La lutte contre la Nouvelle Sorbonne* (Paris: Aux amateurs de livres, 1988).

19 Agathon, *L'esprit de la nouvelle Sorbonne. La crise de la culture classique. La crise du français* (Paris: Mercure de France, 1911 [1910]), p. 93; see Berr, 'Au bout de dix ans'.

20 H.L. Wesseling, 'Commotion at the Sorbonne: The Debate on the French University, 1910–1914', *European Review* 9/1 (2001): 89–96 (89).

21 Berr, 'Au bout de dix ans', pp. 7–8.

22 See Sirinelli, *Génération intellectuelle*, pp. 226–9. Sirinelli points out that the number of ENS students signing petitions promoted by radical and socialist students contradicts Agathon's 'findings'.

became very close) in 1939 published an obituary of Lévy-Bruhl in which they claimed that 'as many Jews of his generation' he had been influenced by German philosophy, and by another Jew, Durkheim. They regarded the 'pseudo-science' of sociology as part of a movement of 'republicanization' and 'secularization' of teaching.[23] It is without doubt that Durkheim, Lévy-Bruhl, Berr and the large majority of the scholars involved in new disciplines were indeed republican and secular, and many of them were socialists. Their interest in science and scientific method was part of their cultural and political outlook.

The Centre de synthèse: Historical Synthesis and the History of Science

In Berr's mind, the *Revue de synthèse*, the Centre de synthèse, and the book series *L'évolution de l'humanité* shared the same aim: the practical realization of the idea of historical synthesis, which he had first formulated in his doctoral dissertation.[24] By 'historical synthesis', Berr meant both a synthesis of all disciplinary points of view, and of as many 'positive facts' as possible; indeed, for him the object of history should not exclude anything on principle. 'Historical synthesis', which encompassed all the partial approaches of current historical research, was his most powerful weapon in his mission to rescue the human sciences from the 'chaos' in which for him they had fallen. He believed that the 'historical method' was the best method in one's search for truth.[25] His views were reflected on the activities of the Centre de synthèse, which were aimed at being very comprehensive both in terms of subject matter and of disciplinary approach. Indeed, the work of preparation of bibliographies and of the historical dictionary that was carried out at the Centre de synthèse, was the result of his conviction that in order to reach historical truth it was necessary to have as complete a view of the past as possible.[26] In his view,

23 This obituary is quoted in Merllié, 'Lévy-Bruhl et Durkheim. Notes biographiques en marge d'une correspondance', p. 494.

24 His doctoral dissertation is Henri Berr, *L'avenir de la philosophie: Esquisse d'une synthèse de connaissances fondée sur l'histoire* (Paris: Hachette, 1899); on *L'évolution de l'humanité*, see Jacqueline Pluet-Despatin, 'Henri Berr éditeur. Elaboration et production de "L'évolution de l'humanité" ', in Biard, Bourel and Brian (eds), *Henri Berr et la culture du XXe siècle*, pp. 241–67.

25 Berr, *L'avenir de la philosophie*, p. 443; Henri Berr, *La synthèse en histoire: Essai critique et théorique* (Paris: Alcan, 1911), p. 232.

26 For an overview of Berr's many activities and their spirit, see Robert Bouvier, 'Henri Berr et son œuvre', *Revue de synthèse* 85 (1964): 39–50; on the early days of the *Revue de synthèse historique* see Martin Fugler, 'Fondateurs et collaborateurs, les débuts de la *Revue de synthèse historique*', in Biard, Bourel and Brian (eds), *Henri Berr et la culture du XXe siècle*, pp. 173–88; on the *Vocabulaire* see Margherita Platania, 'Le projet de *Vocabulaire historique*', in Biard, Bourel and Brian (eds), *Henri Berr et la culture du XXe siècle*, pp. 231–9; and Margherita Platania, *Le parole di Clio: Polemiche storiografiche in Francia 1925–1945* (Napoli: Bibliopolis, 2001). On the Semaines de synthèse, which was a series of conferences organized by the Centre de synthèse, see Bernadette Bensaude-Vincent, 'Présences scientifique aux semaines de synthèse', in Biard, Bourel and Brian (eds), *Henri Berr et la culture du XXe siècle*, pp. 219–30; and Marina Neri, 'Vers une histoire psychologique: Henri Berr et les

the collection of data alone, however, was not sufficient: a work of generalization was the further step towards obtaining general knowledge from positive facts.[27] He proposed an inductive method which would make history a scientific discipline. He was careful to avoid the suggestion that his 'historical synthesis' could have anything to do with the synthesis offered by the philosophy of history, especially as practised in Germany.[28] Berr's polemic against *a priori* systems was attuned to that of such historians of philosophy as Delbos, Bréhier, Brunschvicg and of course his own teacher Boutroux.[29] However, his 'scientific' proposal was very different from their philosophies. For Berr, history, as a science, is neither a metaphysics, which is general but *a priori*, nor erudition [*érudition*], which is a mere collection of facts that does not attain general knowledge. In his view, Germany had produced both erudition and 'adventurous syntheses', while Berr proposed a 'French style' synthesis.[30] The members of the Centre de synthèse by no means all shared Henri Berr's theory of knowledge, but his positive view of science, along with his secularism and progressive politics, formed the common ground for many intellectuals who aimed to be innovators.

In Berr's eyes all the Centre de synthèse's activities were about the sciences, these being the human sciences ('sciences de l'homme', including history, sociology, psychology and ethics) and the natural sciences ('sciences de la nature', including mathematics, physics and biology).[31] For Berr history was as much a science as physics, though not yet as 'advanced' as the latter. As far as the natural sciences were concerned, they were well represented at the Centre de synthèse, and Rey and Langevin played a crucial role in keeping the discussion on scientific issues very much alive.[32] Indeed, in Febvre's view, they almost took over the Centre de synthèse altogether. So much so that he complained first in a letter sent to Bloch in April 1934, and six months later in a letter to Berr himself, about the increasingly large space that Rey and his associates were gaining at the Centre de synthèse.[33]

Semaines internationales de synthèse (1929–1947)', in Biard, Bourel and Brian (eds), *Henri Berr et la culture du XXe siècle*, pp. 205–17.

27 Berr, 'Au bout de dix ans', p. 5; Anonymous [Henri Berr], 'Sur notre programme', *Revue de synthèse historique* 1/1 (1900): 1–8. The article that followed Berr's (unsigned) presentation of the first issue of the *Revue de Synthèse* was written by Boutroux: Emile Boutroux, 'Histoire et synthèse', *Revue de synthèse historique* 1/1 (1900): 9–13.

28 Berr, 'Au bout de dix ans', p. 2; Berr, *La synthèse en histoire*, p. 260.

29 See Chapter 2.

30 Henri Berr, 'Introduction à une Histoire Universelle', *Revue de synthèse historique* (1920): 17–34 (18).

31 Henri Berr, 'Au bout de trente ans. La Revue organe du Centre', *Revue de synthèse* 51 (1931): 3–8 (6). On this point see Bensaude-Vincent, 'Présences scientifique aux semaines de synthèse', p. 219.

32 See Bensaude-Vincent, 'Présences scientifique aux semaines de synthèse', for a qualitative and quantitave analysis of the presence of the sciences at the Semaines de synthèse.

33 Febvre's letter to Bloch is quoted in Müller, 'Lucien Febvre et Henri Berr: de la synthèse à l'histoire-problème', p. 52; his letter to Berr is published in Febvre, *De la Revue de synthèse aux Annales*, p. 510. See Müller's article also for the difficulties that slowly arose

Berr always stressed the importance of the history of science, and not only as one of the necessary components of the historical synthesis, but also because for him the history of science was the history of progress. As a discipline, the history of science was present in the *Revue de synthèse historique* from the very first issues, with articles by Paul Tannery, including the speech 'De l'histoire générale des sciences', which he had prepared as inaugural lecture for the chair at the Collège de France, which, however, he failed to obtain.[34] Both Rey and the members of the Section d'histoire des sciences assured that the history of science played a role in the activities of the Centre de synthèse. However, the influence and personal prestige of Abel Rey was not comparable with that of the Director of the Section d'histoire des sciences Aldo Mieli, and this difference reflected on the importance accorded to their projects. Rey was not only a very close associate of Berr, and the co-editor of the *Revue de Synthèse*, but also a professor at the Sorbonne. He represented the official side of the history of science; the members of the Section de synthèse générale, which he directed, included such famous Sorbonne professors as Lucien Lévy-Bruhl,[35] Bréhier, Brunschvicg, Lalande, Laporte, Robin and the Dean of the Faculty of Letters Henri Delacroix, along with the Collège de France psychologists Pierre Janet and Henri Piéron, and Emile Meyerson, who was not an academic but was nevertheless a famous philosopher.[36]

On the other hand, Aldo Mieli was a foreigner and had no academic position in France. The administrator, treasurer and librarian of the Section d'histoire des sciences, Metzger, never held a salaried academic post, excluding her temporary replacement of Alexandre Koyré at the Ecole pratique des hautes etudes. She also lectured in Rey's institute at the Sorbonne, but she was not paid. Berr, Rey and Mieli were however united in their defence of science, and in their aim to extend the methods of the natural sciences to the human sciences. In the following two sections, I shall discuss Rey's and Mieli's activities and ideas.

Abel Rey between 'Scientism' and Mentalities, and between the Sorbonne and the Centre international de synthèse

Rey's Conception of Philosophy and the History of Science

Abel Rey played a major role in the development of the history of science in France. In organizational terms, his major achievements were the Institut d'histoire des

---

between Berr and Febvre after Febvre founded the *Annales* and was given the editorship of the *Encyclopédie française*.

34 Paul Tannery, 'De l'histoire générale des sciences', *Revue de synthèse historique* 8 (1904): 1–16.

35 Lucien Lévy-Bruhl's son Henri, was member of the Section de synthèse historique; see Fondation 'Pour la science', *Bulletin du Centre International de Synthèse. Section de synthèse historique* (Paris: Appendix of *Revue de synthèse*, 42, nouvelle série 16, 1926), p. 34.

36 See the list of members in Rédaction, 'La vie du Centre. La section de synthèse générale. Son programme', pp. 130–31.

sciences et techniques, and its journal *Thalès*. When he founded the Institut, he was already a very central and active member of the Centre de synthèse: he was Director of the Section de synthèse générale and co-editor of the *Revue de Synthèse*; he also wrote five volumes on ancient science for the *L'évolution de l'humanité* series.[37] For Berr, Rey was 'the' historian of science of the Centre de synthèse. Already in 1912, Berr signalled to Sarton that his 'friend' Abel Rey, who was then Professor of Philosophy at the University of Dijon, could be 'very useful' to Sarton and *Isis*, the history of science journal that he was launching.[38] Berr and Rey shared the belief that scientific knowledge is the only type of valid knowledge, and that scientific methods should be applied to any domain of knowledge. They also shared what one can roughly describe as a positivist and empiricist conception of knowledge.

Particularly in his early works, Rey repeatedly declared his positivism, indeed his 'absolute positivism'.[39] However, Rey distinguished his own 'positivism' from Comte's. It is in order to distinguish his philosophy from Comte's that he stated that, rather than a positivist, he was a '*scientiste*', or a follower of science. His own positivism, or 'scientism', he explained, meant that he only accepted the teachings of positive science.[40] He even argued that Comte was indirectly to blame for the disregard in which in his view contemporary philosophy held science. He argued that Comte, by condemning all reflections on origins, which were the domain of metaphysics, and by condemning metaphysics itself, had indirectly created an 'illusion' that true explanations were to be found outside science. However, some developments of German metaphysics for him had been far more negative than Comte's philosophy: already Kant, by limiting the method of knowledge of the universe to 'Newtonian mathematicism' and excluding metaphysics from it, had in Rey's view opened the possibility of knowledge outside science. Philosophers like Schelling and Schopenhauer wholeheartedly chose the path that Kant had opened. For Rey in this way, anti-intellectualism and mysticism, including eastern mysticism, entered philosophy.[41] There is no doubt that he thought that this was a very unfortunate development; it was not just science that he defended, but also the west and modernity. The 'purely scientific' attitude that he advocated was in his view the legacy of the Renaissance and above all of the Enlightenment.[42]

37 Rey, *La science dans l'antiquité, vol.1: La science orientale avant le grecs* , Rey, *La science dans l'antiquité, vol. 2: La jeunesse de la science grecque*; Rey, *La science dans l'antiquité, vol. 3: La maturité de la pensée scientifique en Grèce*; Rey, *La science dans l'antiquité, vol. 4: L'apogée de la science technique grecque: Les sciences de la nature et de l'homme, les mathématiques d'Hippocrate à Platon*; Rey, *La science dans l'antiquité, vol. 5: L'apogée de la science technique grecque: L'essor de la mathématique* .

38 Henri Berr, Letter to Sarton of 4 May 1912, Houghton Library, bMS Am 1803 (111).

39 'Towards absolute positivism' is the title of one of his articles: Abel Rey, 'Vers le positivisme absolu', *Revue philosophique* 34 (1909): 461–79. See also Abel Rey, 'Sur le positivisme absolu', *Revue philosophique* 34 (1909): 65–6.

40 Abel Rey, *La philosophie moderne* (Paris: Flammarion, 1908), p. 6, n. 1.

41 Ibid., pp. 21–2.

42 Ibid., p. 39.

He was equally critical of contemporary philosophers who regarded the domain of science as limited. Indeed, he argued that Pierre Duhem had proposed a metaphysical interpretation of science, despite his claims to the contrary, because he had imposed *a priori* limits to scientific knowledge. Duhem's philosophy, he concluded, was the philosophy of a believer.[43] Rey did not admit any other knowledge but scientific knowledge, and rejected the possibility of knowledge of objects that are outside human experience. He argued that 'experience is the criterion of truth, and therefore of objectivity'.[44] Despite their philosophical differences, Rey and Brunschvicg agreed that what is outside human experience is just as if it did not exist, as far as human beings are concerned.[45]

If for Rey there is no other knowledge than science, what is for him the role of philosophy? He thought that philosophy and the history of science coincide. As he put it already in 1909, 'the philosopher is the historian of contemporary scientific thought',[46] and he should apply historical methods and criticism to the study of scientific thought.[47] Moreover, in his view, philosophy does not have different objects or methods from science, but only a different level of generality. He thought that philosophy, that is to say the history of science, should make a synthesis of the doctrines produced by the various sciences, and link the sciences and *culture générale*.[48] Rey's conception of philosophy and the history of science differs from that of Brunschvicg. While for Rey philosophy is the history of science, for Brunschvicg philosophy is a reflection on the history of science. For the former, philosophy and science have the same object, while for Brunschvicg the object of philosophy is the history of science. For Brunschvicg philosophy is distinct from science as far as object and method are concerned.

Rey was not satisfied with the use that many philosophers made of the history of science, which in his view was that of a mere repository of examples. Once he remarked that Brunschvicg's history of mathematics was that of a philosopher who employed the history of science in order to prove his theses, rather than that of a

---

43 Abel Rey, 'La philosophie scientifique de M. Duhem', *Revue de métaphysique et de morale* 12 (1904): 699–744 (733, 744). In his famous article 'Physique du croyant' (1905), Duhem replied that his was indeed a believer's physics. Their polemics did not stop there: see the chapter dedicated to Duhem in Rey's *La théorie physique*, originally published in 1907 (Abel Rey, *La théorie de la physique chez les physiciens contemporains. Troisième édition revue et augmentée* [Paris: Alcan, 1930], vol. 2, ch. 3) and Duhem's review of it (1907). Duhem chose to add his two articles (1905 and 1907) to the 1914 edition of *La théorie physique*: Pierre Duhem, *La théorie physique. Son objet, sa structure. Deuxième édition revue et augmentée* (Paris: Vrin, 1981 [1914]). On this controversy between Duhem and Rey, see Redondi, *Epistemologia e storia della scienza*, pp. 33–36; and Jaki, *Uneasy Genius: The Life and the Work of Pierre Duhem*, pp. 341–50.

44 Rey, *La théorie de la physique chez les physiciens contemporains*, p. 282.

45 Abel Rey, 'A propos de "l'explication dans les sciences" de M. Meyerson', *Revue de synthèse historique* 32 (1921): 123–40 (127). Cf. Brunschvicg, *L'idéalisme contemporain*, discussed in Chapter 3 above.

46 Rey, 'Vers le positivisme absolu', p. 472.

47 Rey, *La théorie de la physique chez les physiciens contemporain*, p. 478.

48 Rey, 'Vers le positivisme absolu', p. 473.

historian of science. For Rey, Brunschvicg wanted 'to do philosophy of science' rather than 'history of the sciences'.[49] Similarly, Emile Meyerson, in his study of the mind, had used, in Rey's view, historical examples selectively to support his ideas, which he had elaborated independently of a scientific study, that is to say a study based on 'facts' and evidence.[50] Rey argued that the history of science should in fact be based on 'scientific' empirical work aimed to offer as complete a picture of the past as possible. For him, the first, and more basic task, carried out by the historians of particular sciences, involved the gathering of facts, their description and organization.[51] The second task was aimed at a general history of science, in which the works of the specialized historians were coordinated and given a more general interpretation. Rey argued that both levels of the history of science are necessary and complementary.[52] This was the method of the Centre de synthèse with its extensive programme of bibliographies, vocabularies and in-depth research on small portions of history that would eventually be merged into a general history.

The combination of extensive knowledge of 'historical facts' and of their critical evaluation, was also the guiding idea of Abel Rey's own Institut d'histoire des sciences et techniques. Rey did make sure that his students would know as much history of science as possible. The programme of general history of science was remarkably comprehensive as far as chronology was concerned, as it started with technical and scientific knowledge before the sixth century BCE and ended with present-day science. It also included many sciences: not only mathematics, physics, chemistry, biology and medicine, but also eighteenth-century moral and political sciences in relation to the natural sciences, and, for the more recent periods, psychology and sociology.[53] A student studying towards a *certificat* would have had a choice between two options, one more focused on the 'general and philosophical history of the sciences', and the second on 'history of contemporary scientific theories and of their philosophical interpretations'. The two corresponding general learning outcomes sounded rather

---

49 Abel Rey, 'Revue d'histoire des sciences et d'histoire de la philosophie dans ses rapports avec les sciences', *Revue de synthèse historique* 31 (1920): 121–35 (129).

50 Rey, 'A propos de "l'explication dans les sciences" de M. Meyerson', p. 124. Frédéric Fruteau de Laclos has analyzed Rey's critique of Meyerson in his very comprehensive study of Meyerson: Frédéric Fruteau de Laclos, 'La philosophie de l'intellect d'Emile Meyerson. De l'épistémologie à la psychologie' (unpublished thesis: Université de Paris-X Nanterre, 2004).

51 Abel Rey, 'Histoire de la science ou histoire des sciences', *Archeion* 12/1 (1930): 1–4 (3).

52 Ibid.

53 The Institut d'histoire des science et techniques administrated both a *certificat* and a diploma in history and philosophy of the sciences. The *certificat* counted towards a degree in humanities. The diploma required a thesis, its viva voce discussion and an oral examination on a question chosen from the Institut's programme. See Anonymous, 'Programme détaillé du certificat d'histoire et philosophie des sciences et du diplôme de l'Institut', *Thalès* 3 (1936): 233–47 (247). On the first years of the Institut and its aims, see Jean-François Braunstein, 'Abel Rey et les débuts de l'Institut d'histoire des science et techniques (1932–1940)', in Michel Bitbol and Jean Gayon (eds), *L'épistémologie française, 1830–1970* (Paris: Presses Universitaires de France, 2006), pp. 173–91.

militant. The first outcome, which was 'to encourage the studies of history of the sciences in their relationship with general civilisation', also claimed that the sciences formed a hitherto neglected part of civilization, and that their study would change completely the current conception of history. The learning outcome corresponding to the contemporary option encouraged a study of contemporary science, of its philosophy and of its role in contemporary civilization.[54]

In Rey's view, the new history of science was not only to be defined against a history of science as a repository of examples for the philosopher, but also against the narrow conception of the history of the sciences as collections of facts and discoveries within individual sciences. For him, the latter approach was not only historically inaccurate, but also not attentive to the mentalities behind the theories. On the one hand, he wanted to reform the history of science as practised by his colleagues, and Brunschvicg in particular, on the other hand, he wanted to distinguish his own version of the history of science from that written by such scientists turned historians of science as Pierre Boutroux.[55] When, in 1920, Pierre Boutroux wrote *L'idéal scientifique des mathématiciens*, he contrasted his own work with the most famous recent general work on history of mathematics, Brunschvicg's *Les étapes de la philosophie mathématique*, published eight years earlier. Boutroux presented his own work as 'pure' history of mathematics, as opposed to Brunschvicg's history of scientific and philosophical conceptions. The result of Brunschvicg's approach, Boutroux argued, is a history full of discontinuities and 'oscillations'. Pierre Boutroux claimed that such irregularities disappear if one only examines 'pure' scientific work, completely free from any philosophical concerns.[56] He advocated a separation between the object of the history of science and that of the history of philosophy or other intellectual histories. For him, only the 'specialised scientific workers' and the 'pure technicians' should be considered in a work on the history of science.

Rey did not agree with Boutroux. He conceded that the history of science should be based on the examination of the 'technicians' work', but he added that it should be equally based on the study of the 'great currents of thought' that informed the practitioners' minds. Indeed, Rey's own version of the history of science was a 'history of scientific ideas in its relationship with philosophical doctrines'.[57] He also argued that science and philosophy had been separated only recently, and indeed

54 The two options, as far as the examination was concerned, only differed in one of the two written tests, one being on the general history of science, the other on contemporary sciences. All students had to take a second written test on general philosophy (except those who already held a *certificat* of general philosophy) and two oral tests on general and philosophical history of the sciences, and on contemporary scientific theories and their philosophical interpretations.

55 Pierre Boutroux (1880–1922) was first Professor of Differential Calculus and Mathematics at Poitier and Princeton, and then, in 1920, was elected Professor of the History of the Sciences at the Collège de France.

56 Pierre Boutroux, *L'idéal scientifique des mathématiciens, dans l'antiquité et dans les temps modernes* (Paris: Alcan, 1920), pp. 22–3.

57 Rey, 'Revue d'histoire des sciences et d'histoire de la philosophie dans ses rapports avec les sciences', p. 122.

this separation was rather artificial. It is the historian's duty, he continued, not to introduce a separation where one did not exist.[58]

Outillage mental, *Secularism and Progress*

As he suggested in his response to Pierre Boutroux, Rey was concerned with understanding the minds of the scholars who had authored the doctrines under study. Indeed, in his works on the history of science, he aimed at establishing which type of mentality had produced a certain type of knowledge. For instance, in the first volume of his *Science dans l'antiquité*, he discussed when 'scientific thought' emerged; he also judged the calculations performed by tribal civilizations to be different from modern versions, as the former were not the expression of a scientific mentality.[59]

Rey's most extensive presentation of his conception of the mind and its evolution is to be found in the article on 'the evolution of thought: from primitive thought to current thought' in the *Encyclopédie française*. Rey's article opens the first volume, which was dedicated to the '*outillage mental*' or mental instrument.[60] In Rey's presentation, the concept of *outillage mental* is very close to that of mentality; indeed he extensively cited Lévy-Bruhl. He chose to present the various ways of thinking first synchronically and then diachronically. From a synchronic perspective, the most important opposition was for him that between 'our western thought', on the one hand, and that of 'savages' and young children, on the other.[61] Rey agreed with Lévy-Bruhl that the primitives' 'immediate' and 'intuitive' way of thinking was 'impermeable to critique and experience'. In this, primitive thought was more similar for Rey to that of an 'animal society' than that of the 'discursive', critical and analytical way of science.[62] Rey dismissed Meyerson's criticism of Lévy-Bruhl's 'primitive mentality', according to which there is no profound difference in the mental mechanisms of the 'primitive' and that of the scientist.[63]

58 Ibid., p. 130. Rey's disagreement with Pierre Boutroux mirrors that between Robin on the one hand, and himself and Berr on the other, mentioned in Chapter 2. On that occasion, the historian of philosophy Robin refused to consider the history of philosophy and science together.

59 Rey, *La science dans l'antiquité, vol.1*, p. 431 and passim.

60 The plan of the *Encyclopédie française* was focused on the *outillage* rather than the simple accumulation of information, see Lucien Febvre, 'Une encyclopédie française: pourquoi, comment?', in Lucien Febvre (ed.), *Encyclopédie française* (Paris: Société de Gestion de l' Encyclopédie Française, 1937), p. 1.04.12.

61 Rey put the term 'savages' in inverted commas himself. Elsewhere, he acknowledged the inadequacy of such adjectives as 'primitive', 'inferior' and 'non-civilised' when employed to described societies: Rey, *La science dans l'antiquité, vol. 1*, p. 29.

62 Abel Rey, 'L'évolution de la pensée: de la pensée primitive à la pensée actuelle', in Febvre (ed.), *Encyclopédie française*, p. 1.10.3.

63 For Meyerson's criticism of Lévy-Bruhl, see Emile Meyerson, *Du cheminement de la pensée* (Paris: Alcan, 1931), vol. 1, pp. 81–8. See also the meeting of the Société française de philosophie in which Levy-Bruhl presented his *L'âme primitive* (Lucien Lévy-Bruhl, *L'âme primitive* [Paris: Presses Universitaires de France, 1996 (1927)]): Lévy-Bruhl et al., 'L'âme primitive. Séance du 1er juin 1929'. Meyerson referred to this meeting at the beginning of his

Rey argued that the differences between western and eastern ways of thinking may look less marked than those between modern and primitive mentalities, but they still exist. He regarded western and eastern mentalities as having evolved in different directions from primitive thought: while western thought had opposed primitive thought in its main characteristics, eastern thought for him had prolonged its most important aspects, such as the little attention paid to precision, clarity and distinction of ideas and, in general, to logic. As a consequence, Rey argued, while western thought is oriented towards intellectualism and rationalism, and is directed to the study of the world, eastern thought is impregnated by mysticism, sentiment and religion, and is directed towards the interiority.[64] Rey even found differences in thinking habits among European cultures: he believed that the modern Latin countries, and France in particular, were the direct inheritors of the Greek equilibrium between logical thinking and well-defined representations on the one hand, and use of the intuition on the other, although, he added, the use of intuition should always be provisional, as it must be placed into a logical structure. By contrast, he thought that in Germany the use of intuition had not been contained within logical constraints, despite the work of such thinkers as Kant, Lessing, Goethe and Schiller. Bottom of Rey's list were the Slavic nations, which for him were closest, among Europeans, to eastern thought.[65]

It is however in the diachronic presentation of the history of the mind, or rather of minds, that Rey disagreed with Lévy-Bruhl. Despite the opposition between primitive and modern ways of thinking, Rey still thought that the latter had evolved from the former. Indeed, he presented a six-stage model of the history of the mind, from primitive to current western thought. Rey repeatedly described the move to a new stage as a 'liberation': Greek thought (third stage) was a liberation from myth, modern science (fifth level) was the 'thought's greatest liberation', and current thought (sixth level) a 'new liberation'. 'Liberation' for him is liberation from myth, religion and from any dogma. Indeed, the latest liberation was for Rey from the 'absolutes' of mathematical physics. One of the main differences between Rey's and Comte's models is that the former is open-ended: Rey emphasized that the present '*outillage mental*' was only provisional.[66]

Rey's model of the history of the mind is the mind's progressive liberation from dogma and its increasingly logical and rational approach to its object. Bearing in mind that for Rey modern science is the model of rationality, his outline of intellectual progress consists of progress toward modern science, and from now on, towards an increasingly more scientific approach to all domains of knowledge. Rey regarded the progress from primitive thought to modern thought as a movement from religious to scientific explanation of phenomena. He regarded those philosophies that distinguish

---

discussion of Lévy-Bruhl in *Du cheminement de la pensée*. He could not attend the meeting, but sent a letter. See also Joseph LaLumia, *The Ways of Reason: A Critical Study of the Ideas of Emile Meyerson* (London: Allen & Unwin, 1966), pp. 54–5.

64 Rey, 'L'évolution de la pensée: De la pensée primitive à la pensée actuelle', pp. 1.10.3–1.10.4.

65 Ibid., pp. 1.20.9–1.20.10.

66 Ibid., p. 1.20.11.

scientific knowledge and other types of knowledge, or that limit science to a particular domain, as Trojan horses for religion. Despite their philosophical differences, Rey, Brunschvicg and Lévy-Bruhl shared their secularism and the idea of modernity as scientific progress and the erosion of superstition and religion.[67]

Aldo Mieli and the Creation of National and International Spaces for the History of Science

Aldo Mieli, the Promoter of the History of Science

Unlike the other units of the Centre de synthèse, the Section d'histoire des sciences was not directed by a member of the Centre de synthèse's inner group; it also had its own journal, conferences and international connections. Rather than a unit sprung from the original body of the Centre, the Section d'histoire des sciences was almost a small organization within it. Its director, Aldo Mieli, was similar to Henri Berr in his relentless organizational efforts, economic independence and loose relationship with universities. Mieli's life, however, appears to have been rather more eventful than Berr's, as he moved from Italy to France in 1928, and in 1939 from France to Argentina, where he died in 1950. Born in Livorno (Italy) in 1879 of a wealthy Jewish family, after studying chemistry to receive his '*libera docenza*',[68] he became increasingly interested in its history, to the point of obtaining a second '*libera docenza*' in history of science.[69] From the early 1910s onwards, his activities steadily increased in number and importance; these included both original publications and editions of dictionaries and book series.[70] In 1919 he founded the journal *Archivio*

67 Brunschvicg, Lévy-Bruhl and Rey also had in common their non-intellectual family background. Rey was the son of a merchant, see Charle, *Les professeurs de la Faculté des Lettres de Paris*, p. 184.

68 '*Libera docenza*' was generally obtained by examination and allowed the recipient to hold special courses in universities. The award of '*libera docenza*' acknowledged the candidate's eminence in a particular field, but did not carry a full-time post at a university.

69 Aldo Mieli and Pierre Brunet, *Histoire des sciences. Antiquité* (Paris: Payot, 1935), p. 8; Claudio Pogliano, 'Aldo Mieli, storico della scienza (1879–1950)', *Belfagor* 38 (1983): 537–57 (538). Between 1919 and 1928 Mieli held courses in history of chemistry at the University of Rome and in 1926 at the University of Perugia. See the first lecture of the Perugia course in *Archeion*: Aldo Mieli, 'La storia della scienza in Italia (Prolusione ad un corso di Storia delle scienze, tenuta nella R. Università di Perugia il giorno 9 marzo 1926)', *Archeion* 7 (1926): 36–48.

70 His numerous publications included a history of chemistry (Aldo Mieli, *Pagine di storia della chimica* [Rome: Leonardo da Vinci, 1922]) and the first volume of a general history of science, covering antiquity, which never saw the planned following volumes, but was re-written as *Histoire des sciences: Antiquité*, in collaboration with Pierre Brunet ten years later: Aldo Mieli, *Manuale di storia della scienza. Storia. Antologia. Bibliografia. – Antichità* (Rome: Leonardo da Vinci, 1925). He also published a Spanish-language version of this first volume: Aldo Mieli, *Panorama general de historia de la ciencia* (Buenos Aires: Espasa-Calpe, 1945). About Mieli's plans for a monumental history of science, and his life and works, see Pierre Sergescu, 'Aldo Mieli', *Actes du Congrès International d'Histoire des*

di storia della scienza, subsequently known as *Archeion*, which from 1928 became the official publication of the Comité international d'histoire des sciences, and from 1930 of the Centre de synthèse's Section d'histoire des sciences.[71] The international character of this journal developed gradually. Indeed, in the first issue, Mieli presented it as aimed at promoting the history of science in Italy, and at showing Italy's contributions to science: on the one hand he invited articles on Italian subjects and on the other especially encouraged Italian authors. *Archivio* did start off with somewhat nationalistic intents, reminiscent, however, of Risorgimento rather than the then emergent Fascist nationalism. In fact, Mieli was a socialist and never approved of Fascism, whose regime he fled a few years later.[72] In fact, the journal grew more and more international, not just in its topics and authors, but also in its spirit, including the adoption, from 1927, of the artificial language 'Interlingua' (also called 'Latino sine flexione') for its abstracts.[73] Mieli also created a large network of

---

Sciences 6 (1951): 79–95. Mieli also contributed to the Italian journals *Rivista di filosofia* and *Scientia*, and then to *Isis* when it was launched; he became the editor of a series of classic works in science and philosophy, and of a series on Italian scientists; on these activities, see Antonio Di Meo, 'Aldo Mieli e la storia della chimica in Italia', *Archives internationales d'histoire des sciences* 36 /117 (1986): 337–61 (352); José Babini, 'Para una bibliografia de Aldo Mieli', *Physis* 21 (1979): 357–424 (420); Massimo Bucciantini, 'George Sarton e Aldo Mieli: bibliografia e concezioni della scienza a confronto', *Nuncius. Annali di storia della scienza* 2/2 (1987): 229–39 (232). He edited a dictionary of Italian scientists, although only the first volume was actually published with a total of about 60 entries: Aldo Mieli (ed.), *Gli scienziati italiani dall'inizio del Medioevo ai nostri giorni* (Rome: Nardecchia, 1921).

71 After the Second World War, its name was changed to the current *Archives internationales d'histoire des sciences.*

72 Aldo Mieli, 'Digressions autobiographiques sous forme de préface à un panorama général d'Histoire des Sciences', *Archeion* 37 (1947–48): 494–505 (503). He defined himself an 'evangelical socialist' close to Camillo Prampolini. Camillo Prampolini (1859–1930), Italian humanitarian and reformist socialist, organized several cooperatives and was first elected member of the Italian parliament in 1890. One of Prampolini's best-known speeches is the so-called 'Christmas speech': Camillo Prampolini, *La predica di Natale* (Milan: Fondazione Giangiacomo Feltrinelli, n.d. [1897]). Before the First World War, Mieli wrote for a local socialist paper (*Il lavoratore*): Pogliano, 'Aldo Mieli, storico della scienza (1879–1950)', p. 546. He expressed his pacifist beliefs in the words with which he dedicated his *Le scuole ionica pythagorica ed eleata* to George Sarton 'to the Belgian fellow of studies, George Sarton, for his heroic fatherland, for his painful exile, reaffirming on these pages of serene studies the condemnation of the ancient [and] renewed folly: war' Aldo Mieli, *Le scuole ionica, pythagorica ed eleata (I prearistotelici, I.)* (Firenze: Libreria della Voce, 1916). (Note that Mieli employed his own Italian transcription of Greek words, hence 'pythagorica' rather than 'pitagorica', see Mieli, *Le scuole ionica, pythagorica ed eleata (I prearistotelici, I.)*, pp. xv–xvi). Sarton had left Belgium after the German army had occupied Ghent, where he was living with his wife and daughter, see Lewis Pyenson, 'Inventory as a Route to Understanding: Sarton, Neugebauer, and Sources', *History of Science* 33 (1995): 253–82 (257).

73 The new title *Archeion* was the translation of *Archivio.* Mieli had corresponded with Giuseppe Peano, the mathematician who had invented Interligua: Aldo Mieli, '[untitled]' *Archeion* 8 (1927): 1–5 (4). On Interlingua, see Mario Gliozzi, 'Appendice ad Praefatione de volume XXL', *Archeion* 21 (1938): 10–12; and Giuseppe Peano, *Exemplo de Interlingua*

consulting editors in many countries; for France, these were Abel Rey and Hélène Metzger.

Mieli's collaboration with Rey and especially with Metzger grew much closer when he moved to France. His reasons for his removal lay in the difficulties he had due to the Fascist regime. As he explained in a long letter to Sarton 'the gang of criminals who rule[d] Italy', and who had already eliminated their political rivals through murders, convictions and internal exile, were now spying on everybody.[74] It was quite clear to him that the history of science could not flourish in Italy, and indicated two reasons to Sarton: firstly, the new official 'fascist science' was imposed, secondly, the regime had made an alliance with the Vatican; as a consequence the Jesuits had greatly increased their influence.[75] Mieli even feared that the Jesuits' influence could go as far as making the teaching of Darwinism illegal.[76] His politics, conception of science and very likely his sexuality made his life in Italy extremely difficult, and presumably dangerous.[77]

Both Mieli and the Centre de synthèse profited from the former's move to Paris. Mieli found refuge and the possibility of promoting the history of science freely. The Centre de synthèse acquired Mieli's library, and above all an established journal

---

cum Vocabulario Interlingua Latino-Italiano-Français-Deutsch (Cavoretto and Torino: Interlingua, 1913).

74 Mieli's letter to Sarton, 4 November 1927, Houghton Library, bMs Am 1803 (1039). Mieli was right about the Fascist authorities' spying: they kept a dossier on him, and from 1929, when his name was discovered in the address book of an anarchic activist, he was regularly followed: see Carola Susani, 'Una critica della norma nell'Italia del fascismo', in Enrico Venturelli (ed.), *Le parole e la storia: Ricerche su omosessualità e cultura* (Bologna: Il Cassero, 1991), pp. 110–19 (116, n. 2). I thank Enrico Venturelli for kindly sending me her hard-to-find article.

75 The alliance between the Fascist regime and the Vatican that Mieli mentioned was going to be ratified in 1929 with a concordat between the Italian State and the Vatican ('*Patti lateranensi*'). The Italian state thus far had had very difficult relations with the Vatican: between 1874 and 1919 the Vatican even forbade Italian Catholics to take part in Italian politics.

76 Aldo Mieli's letter to Sarton, 4 November 1927, Houghton Library, bMs Am 1803 (1039).

77 Aldo Mieli has become an important figure in Italian gay culture, and websites like Wikipedia (Italian version) call him a 'pioneer' of the gay liberation movement in Italy (see Contributori di Wikipedia, 'Aldo Mieli' http://it.wikipedia.org/wiki/Aldo_Mieli [accessed 30 April 2007]). It is without doubt that he was interested in homosexuality at least in his research, to which he gave ample space in his journal *Rassegna di studi sessuali*. See also his article Aldo Mieli, 'Patologia sessuale', *Archeion* 1/2 (1927): 81–94. In 1921, he was invited to the International Congress of Sexuology in Berlin. Of this trip, he recalled in particular the visits to theatres and clubs for homosexual men and women: see Aldo Mieli, 'Un viaggio in Germania', *Archivio di storia della scienza* 7 (1926): 343–81 (345). Carola Susani reports that Mieli was expelled in 1902 from a Socialist circle for being a 'pederast', and the Fascist authorities noted in his dossier that his relationship with his two flatmates in Paris was connected with his 'sexual abnormality': Susani, 'Una critica della norma nell'Italia del fascismo', p. 116, n. 2.

specialized in the history of science.[78] In the letter to Sarton cited above, Mieli also presented his project of an international committee dedicated to the history of science to be formed at the Oslo sixth international conference of historical sciences the following year. In Oslo, Mieli officially proposed the formation of the Comité international d'histoire des sciences, and a group of scholars was nominated in order to co-opt the first members. This provisional management committee, chaired by Mieli, included Henri Berr, Abel Rey, George Sarton and Charles Singer.[79] The first meeting of the Comité international d'histoire des sciences took then place in its headquarters at the Hôtel de Nevers, in the rooms of the Centre de synthèse. Hélène Metzger was nominated as member of the Comité the following December,[80] Léon Brunschvicg as correspondent member in 1929, and so was Emile Meyerson in 1930.[81]

With Mieli's arrival at the Centre de synthèse, the history of science received a tremendous impulse; from there three organizations were managed: the Section d'histoire des sciences, the Comité international d'histoire des sciences, and the journal *Archeion*. The Section ran regular seminars, with a similar format to those of the Société française de philosophie, the minutes of which were duly published in *Archeion*. The Comité international d'histoire des sciences organized international conferences, the first of which was held in Paris in 1929. Moreover, the history of science library was not only being expanded, but a catalogue was compiled. The latter was rather labour-intensive, as its cards were designed not only to display the items' information, modelled on the Bibliothèque nationale format, but also to have appended cards with 'analytical and critical' reports.[82] This work was expected of Hélène Metzger, as librarian, although she did not enjoy it and thought that she had no relevant expertise.[83]

In terms of organization of activities, the director of the Section d'histoire des sciences could have been as strong a player as Berr, but, instead of finding political

---

78 Mieli's 1864 books and 2280 pamphlets became the core of the Section's library, and Mieli in exchange received a monthly salary: see Lucia Tosi, 'La trayectoria de Aldo Mieli en el Centre international de synthèse', *Saber y Tiempo* 4 (1997): 449–62 (453–4).

79 Aldo Mieli, 'Pour l'organisation des historiens des sciences', *Archeion* 9 (1928): 497–503. Corsini thinks that the Comité international d'histoire des sciences was Mieli's most distinguished achievement: Andrea Corsini, 'Aldo Mieli', *Rivista di storia delle scienze mediche e naturali* 41/1 (1950): 111–13 (112).

80 *Archeion* 9 (1928): 506.

81 Comité International et Centre International d'Histoire des Sciences, 'Listes des membres du Comité International d'Histoire des Sciences', *Archeion* 11 (1929): 84–86 (85); Comité International et Centre International d'Histoire des Sciences, 'Election de nouveaux membres', *Archeion* 12 (1930): 45.

82 Centre International de Synthèse, 'Le répertoire méthodique de synthèse scientifique', *Archeion* 9 (1928): 510–11 (511).

83 In her letter to Sarton of 2 June 1932, Metzger wrote that she had been reproached for not wholly dedicating herself to the catalogue (Houghton Library, bMS Am 1803 [1032]). The work on the library catalogue reportedly experienced a halt when she died, see Friederich Simon Bodenheimer's letter to Sarton of 11 November 1950 (Houghton Library, bMS Am 1803 [159]).

allies in the government of his country, as Berr did, Mieli had to flee Italy, and in 1939 Europe altogether. On the one hand, Mieli had created his own journal, the Comité international d'histoire des sciences, and a wide international network. The long list of governments and organizations that financially backed the Comité international d'histoire des sciences alone would show Mieli's ability to secure funding for his enterprises. On the other hand, his links with France were not organic; he had no academic position, nor was he part of any social networks. Paris was only a good place from which to manage his activities, including his journal, whose publisher remained the Italian Leonardo da Vinci. His social position there afforded him great intellectual stimuli, but little power.

Different Views of the History of Science

Programmatically, the history of science was an extremely important discipline within the Centre de synthèse. However, the Section d'histoire des sciences appeared to be relatively independent of the central organization, and attracted less important members than the other units. Redondi has suggested that the relative autonomy of the Section d'histoire des sciences was also due to an epistemological difference: while Mieli and Brunet aimed to do history of the sciences, and focused on attributions of discoveries, Rey, Metzger, Febvre, and Koyré later on, aimed to write a philosophical history of science.[84] This view seems easily to explain the role that the Section d'histoire des sciences appeared to play in the economy of the Centre de synthèse: it was in charge of collecting data and bibliographic material, and providing works on very specific topics, while the 'historical synthesis' was to be realized by scholars like Rey.

Redondi is right to distinguish these different approaches to the history of science. However, his distinction should not be interpreted simply as an opposition between the history of the individual sciences and general history of science. In fact, Mieli did not oppose a general history of science; indeed, he defended it rather vigorously. He did not believe that the boundaries between disciplines are either clear-cut or stable through history; he argued that they are in fact continuously re-negotiated.[85] The histories of the individual sciences may study specific problems, but for him it is the general history of science that should represent the historical development of science and the mutual relationships among the sciences.[86] He criticized the mathematician and historian of mathematics Gino Loria, who had judged the Collège de France chair of general history of the sciences a rather absurd institution, and who had argued that the gaps in knowledge in the history of the specific sciences were so

---

84 Redondi in Alexandre Koyré, *De la mystique à la science. Cours, conférences et documents 1922–1962, édité par Pietro Redondi* (Paris: Ed. de l'Ecole des Hautes Etudes, 1986), p. xv; see also Tosi, 'La trayectoria de Aldo Mieli en el Centre international de synthèse', p. 458.

85 Aldo Mieli, 'Herbertz Richard, *Philosophie und Einzelwissenschaften*', *Rivista di filosofia* 5 (1913): 300–302 (301).

86 Aldo Mieli, *La storia della scienza in Italia. Saggio di bibliografia di storia della scienza* (Firenze: Libreria della voce, 1916), p. 6.

numerous that historians should first aim to fill them and then maybe attempt a synthesis. Mieli responded that if one were to follow Loria's advice, the synthesis would have never been achieved, as analysis and synthesis proceed hand-in-hand by mutual adjustment.[87]

It is hardly necessary to point out the similarity of Mieli's claims with Berr's much more complex defence of historical synthesis. The similarities between Mieli's and Berr's positions did not end there. First, like Berr, Mieli rejected any distinction between the history of science and the history of philosophy, and argued that science and philosophy have been indistinguishable in many historical periods, such as the Christian Middle Ages and the Renaissance, and in fact they had never been really separate, even when they have been conceived as such.[88] Indeed, he argued that philosophy was one of the sciences, especially as theory of knowledge and theory of values.[89] In this, Mieli was in agreement not only with Berr, but also with Rey, who thought that philosophy must not be a 'literary genre'.[90] Similarly, Mieli thought that history, which for him had started as an art, had recently become a science, and it should be practised in a scientific manner. This was once again in line with the spirit of the Centre de synthèse as a whole.

Mieli and Berr were also very close in their drive to collect all available data and bibliographic information, in order to make history as 'objective' as possible. They shared this particular sense of objectivity, which was supposed to result from the comprehensiveness of information. Mieli aimed to offer historians the tools of their trade, as was evident already in the books that he published in the Italian period of his life: it would be enough to glance at the ample bibliographies of *La scienza greca*, which are duly annotated with Mieli's views. The directorship of the Section d'histoire des sciences suited him as far as its mission to provide an exhaustive repository of the historian's tools. Bibliographies were one of Mieli's more pressing preoccupations, and in this his constant term of reference was George Sarton, with whom he corresponded all his life. However, their conceptions of what a history of science bibliography should include did not overlap completely. Mieli thought that Sarton's bibliographies were not focused enough.[91] He wanted to limit his bibliographies to the history of those disciplines that for him counted as sciences; these included the history of philosophy, but excluded many other disciplines that appeared in Sarton's bibliographies, such as archaeology, anthropology, ethnology and the history of religions.[92]

---

87 Aldo Mieli, 'Sul concetto di storia della scienza', *Rivista di storia critica delle scienze mediche e naturali* 7 (1916): 42–46 (44).

88 Mieli, *La storia della scienza in Italia*, p. 5; Mieli, 'Herbertz Richard, *Philosophie und Einzelwissenschaften*', p. 301; Aldo Mieli, 'Storia delle scienze; rassegna e note sulle pubblicazioni ad essa relative', *Rivista di filosofia* 4 (1912): 508–21.

89 Aldo Mieli, 'Scienza e filosofia', *Rivista di filosofia* 2/5 (1910): 599–608 (608).

90 Rey, *La philosophie moderne*, p. 23.

91 Aldo Mieli, 'Bibliografia metodica dei lavori di storia della scienza pubblicati in Italia', *Archivio di storia della scienza* 1 (1919): 84–86 (85, n. 1).

92 Massimo Bucciantini has analysed the differences between Sarton's and Mieli's conceptions of bibliography: Bucciantini, 'George Sarton e Aldo Mieli: bibliografia e concezioni della scienza a confronto'.

Mieli's view of bibliography corresponded to his substantially internalist practice of the history of science. In the Introduction of their *Histoire des sciences*, Mieli and Brunet paid lip service to the importance of the 'political, economic and cultural context' of the history of science, and even wrote that they aimed to show the milieus and the ways of thinking of the ages under study.[93] This declaration may look close to what our historians of the mind intended to do, but in fact it is not. Mieli and Brunet intended to achieve that objective simply by discussing as many scientists as possible, and by adding biographical notes and a choice of original sources.[94]

Mieli's view of which disciplines should be included in the history of science also reveals a crucial difference with Berr and his other colleagues at the Centre de synthèse. Mieli's brand of 'historical synthesis' was limited to the history of science, which he defined as the study of 'the development' of all 'the activities of the human thought' aimed at 'the systematic knowledge of reality, either speculatively or empirically'.[95] Not only was he not interested in disciplines, such as the history of art, the object of which he regarded as based on 'sentiment' rather than rationality, but, more controversially, he also excluded the social sciences from the history of science. His positivism had taken shape in Italy and partly in Germany, and the French social sciences had not been part of it.[96] The rational analysis of religion and of 'non-rational' thought was outside his cultural domain. His lack of interest in the study of mentalities and in economic and political history could hardly have brought him close to historians such as Febvre, and to those historians of science who aimed to analyse the mind. For Mieli the history of science was not a history of the mind, but rather a history of facts, discoveries and theories. This created a noticeable difference between himself on the one hand, and Rey, Metzger and Koyré on the other.

Despite the views of its director, it is in the Section d'histoire des sciences that important discussions on the history of science as the history of the mind took place. The two most important historians of science in this respect to be met there were its administrator, Hélène Metzger, and Alexandre Koyré. On paper, Mieli and Metzger had much in common: they were both chemists turned historians of science, they were both from well-off Jewish families, and both had socialist inclinations. They collaborated for many years in the running of the Section d'histoire des sciences; Mieli's greetings added to her letters to Sarton and Charles and Dorothea Singer witness to their sharing of spaces and times.[97] However, they never collaborated on a publication, and their relationship appeared to be strangely asymmetrical. In terms of power within the Section d'histoire des sciences, and the Comité international d'histoire des sciences, it was quite clear that Mieli played a far more important

93 Mieli and Brunet, *Histoire des sciences. Antiquité*, p. 9.

94 Ibid., p. 18.

95 Mieli, *La storia della scienza in Italia*, pp. 3–4.

96 After his degree in chemistry, Mieli spent a semester in Leipzig, where he studied with Wilhelm Ostwald, but also read Ernst Mach's books, which, as Mieli wrote, were crucial in shaping his future research, see Mieli, 'Un viaggio in Germania', pp. 343–4.

97 Metzger's letters to Charles and Dorothea Singer are kept in the Wellcome Institute Library, London: PP/CJS.

role; he had created these organizations and made most of the decisions. In public speeches and reports, he hardly ever mentioned that 'enthusiastic and hard-working woman', as he described her later in life,[98] he did not inform her of his intention to move to Argentina, and, once there, he did not reply to her letters.[99] On the other hand, he wrote very positive reviews of her books, even before moving to France,[100] while Metzger privately expressed serious reservations on his work.[101] It is quite clear that their scholarly aims were quite different; Metzger's interest in ways of thinking, and her aim to understand the minds of past scholars must have looked to Mieli like diversions from the real work.

---

98 Mieli, 'Digressions autobiographiques sous forme de préface à un panorama général d'Histoire des Sciences', p. 499.

99 Metzger's letter to Sarton, 26 June 1940 (Houghton Library, bMS AM 1803 [1032]). By contrast, Mieli informed Sarton of his plans: writing from Leiden, on 18 May 1939, Mieli informed Sarton that he had obtained his Argentinian visa and was planning to leave Europe from Antwerp on 30 May (Mieli, Letter to Sarton of 18 May 1939, Houghton Library, bMs Am1803 [1039]). Mieli, who had fled Fascist Italy, had to experience the rise of another dictatorship in Argentina. He described the Argentine military dictatorship as 'corrupt, fraudulent, anti-democratic, clerical and also anti-Semite' in a letter to Sigerist, undated, but written after 8 September 1943, as he refers to the fall of Fascism, but before the end of the war (Yale University Library, Sigerist Papers, Box 4, f. 206).

100 Aldo Mieli, 'Hélène Metzger, *La genèse de la sciences des cristaux; Les doctrines chimiques en France du début du XVII à la fin du XVII siècle*', *Archivio di storia della scienza* 4 (1923); 290–91; Aldo Mieli, 'Hélène Metzger, *Les concepts scientifiques*', *Archeion* 8 (1927): 270–71.

101 Metzger judged Mieli's book on Arabic science (Aldo Mieli, *La science arabe et son rôle dans l'évolution scientifique mondiale. Avec quelques additions de Henri-Paul-Joseph Renaud, Max Meyerhof, Julius Ruska* [Leiden: Brill, 1938]) a 'good handbook' for secondary schools; see Metzger, Letter to Sarton of 5 August 1939, Houghton Library, bMS Am 1803 (1032).

Chapter 5

Approaches to the History of the Mind: The History of Science and the History of Thought

Hélène Metzger and the History of the Mind

A Double Marginality

Hélène Metzger always remained at the margins of academia, despite the value of her publications and her numerous scholarly activities. The daughter of a Jewish precious-stone merchant, she was not expected to pursue an academic career, but rather to marry and raise a family. As discussed in Chapter 1, the marriage pattern of relatively poor Jewish intellectuals to wealthy Jewish women has been sociologically analysed. Metzger clearly knew this pattern from experience: she explained that her father did not allow his 'socialist, or almost' daughters to have independent professions, but rather provided them with dowries, in order to enable them to marry young men 'of intellectual and moral value, but modest means'.[1] Indeed, in 1913 Hélène Bruhl married the historian Paul Metzger who lectured at the University of Lyon. However, he died in the First World War, leaving her a widow with no children, and with her own ample financial means.[2] Similarly to her husband, who had turned to legal history after studying law, Hélène, who had a *diplôme d'études supérieures* in crystallography, developed an interest in the history of science.[3]

---

1 Metzger's letter to Sarton, 22 April 26, Houghton Library, bMS Am 1803 (1032). By contrast, Hélène's half-brother Adrien Bruhl attended the Ecole normale supérieure and eventually became the Dean of the Faculty of Letters at the University of Dijon. In his obituary of Henri Berr, he remembered his sister who worked alongside Mieli and Brunet in Berr's centre, and Berr's friendship towards her: Adrien Bruhl, '[Hommage à Henri Berr]', *Revue de synthèse* 85 (1964): 54. After Hélène's death, he corresponded with Sarton, exchanging documents and Metzger's and their own books; see Adrien Bruhl's letters to Sarton of 17 August 1954, 15 October 1955 and 25 November 1955, Houghton Library, bMS Am 1803 (216).

2 For biographical information, I rely on Gad Freudenthal, 'Hélène Metzger: Eléments de biographie', in Gad Freudenthal (ed.), *Etudes sur/Studies on Hélène Metzger* (Leiden: Brill, 1990) and on Metzger's letters in the Houghton Library, Harvard. Extracts of her letters have been published as an appendix in Freudenthal (ed.), *Etudes sur/ Studies on Hélène Metzger*.

3 Paul and Hélène also shared an interest in the eighteenth century, see Paul Metzger, *Le conseil supérieur et le grand bailliage de Lyon* (Lyon: Université, 1913).

During the war she wrote *La genèse de la science des cristaux*,[4] but soon realized that her science teachers were not interested in the history of science. It was among the philosophers that she found encouragement and support, first from Gaston Milhaud, who however died in 1918, and subsequently from André Lalande, who presented her most philosophical book, *Les concepts scientifiques*, for the Bourdin prize, which she won.[5] She was determined to become a recognized member of the academic community; indeed in 1926, the year after she had been awarded the prize, she wrote to Sarton: 'I shall obtain a real post somewhere. But I am forever stopped in my efforts'.[6] The obstacles to her success were of various types, although linked to one another. The most practical one was that she lacked the right qualifications to obtain a post in the large majority of higher-education institutions.[7] Moreover, women professors were hardly the norm: in 1930, in the whole of France there were only six women holding university positions. At the University of Paris there were only two female professors, both in the Science Faculty, one of whom was the Nobel-prize winner Madame Curie,[8] who had taken over her husband's chair after his death.[9] Metzger's gender played an important role in making her a perennial junior member of her intellectual community. Her difficult relationship with the philosopher Emile Meyerson is a case in point. His 'influence' on her is often mentioned and sometimes discussed.[10] Metzger resented what she perceived as Meyerson's lack of respect for

4 Hélène Metzger, *La genèse de la science des cristaux* (Paris: Alcan, 1918).

5 Hélène Metzger, *Les concepts scientifiques* (Paris: Alcan, 1926). An extract of Lalande's speech at the Académie des sciences morales et politiques, which awarded the prize, is published as Introduction to the volume (pp. vii–ix). He had already presented her *La génése de la science des cristaux* to an English-speaking readership as an important contribution to philosophy, despite some reservations that suggest that for Lalande the author of the book had not still reached philosophical maturity: André Lalande, 'Philosophy in France, 1918', *The Philosophical Review* 28/5 (1919): 443–65 (460–61). It was Rey who presented Metzger's *Les concepts scientifique* to an English-speaking readership, although almost entirely by quoting Lalande: see Abel Rey, 'French Philosophy in 1926 and 1927', *The Philosophical Review* 37/6 (1928): 527–56 (533–4).

6 Metzger, Letter to Sarton, 22 April 1926, Houghton Library, bMS Am 1803 (1032).

7 She gained a *brevet supérieur* rather than a *baccalauréat*, which was a necessary qualification in order to be admitted to a university course ending in a *licence*. She nevertheless went into higher education, eventually obtaining a *diplôme d'études supérieures* in crystallography. Without a *licence*, though, it was not possible for her to go on to a doctorat d'Etat; as a consequence she gained the less prestigious doctorat d'Université, mainly pursued by foreigners or women, who did not aim at an academic career. However, not all academic institutions required formal qualifications; notably, posts at the Ecole pratique des hautes études and the very prestigious Collège de France were assigned purely on scholarly merit.

8 Frances I. Clark, *The Position of Women in Contemporary France* (London: P.S. King & Son, 1937), p. 48. For the complete list of women holding university posts in France in 1930, see E. Charrier, *Evolution intellectuelle féminine* (Paris: Mechelinck, 1931), pp. 407–408.

9 Sarah Dry, *Curie* (London: Haus, 2003), p. 73.

10 Heidelberger and Golinski have argued in favour of his influence, but nevertheless pointed out the differences between their philosophies: see Michael Heidelberger, 'History of Science and Criticism of Positivism: Emile Meyerson's and Hélène Metzger's Views from

her achievements, his expectations of secretarial work from her and his superior attitude towards her. She expressed her impatience and indeed resentment in letters to Sarton, and in one dramatic letter to Meyerson himself, who was ill and would die later that year, 1933.[11] In 1927, she described to Sarton a difficult meeting with Meyerson, who, she wrote, ignored her 'little book', that is *Les concepts scientifiques*, and demanded that she should compile an index for his next book. She wrote that she could proclaim herself to be his disciple if he wished her to do so, but in fact all her publications had been written 'outside his influence'.[12] Metzger thought that her independence from Meyerson should also be made public, and declared it in her paper on his philosophy which she delivered at the first conference organized by the Comité international d'histoire des sciences held in Paris in 1929. She started off by saying that 'it was her pleasure to pay a personal homage' to Meyerson, who 'unfortunately' had not been 'her teacher', as she had read his books '*after* carrying out a great deal of work and publishing two volumes'.[13]

She later made her feelings known directly to Meyerson in a very dramatic letter, which was probably her last to him, as he died seven months after receiving it. She asked him not to treat her as if she were a child, for she was already 43, and argued that 'in the Republic of the minds, we are all equal and you must prove that reason is on your side, not to impose it by force or by intimidation'.[14]

---

a Present-day Perspective', in Freudenthal (ed.), *Etudes sur/Studies on Hélène Metzger*, pp. 151–60; J.V. Golinski, 'Hélène Metzger and the Interpretation of Seventeenth Century Chemistry', *History of Science* 25/1 (1987): 85–97. For an interpretation that emphasizes their differences, see Cristina Chimisso and Gad Freudenthal, 'A Mind of Her Own: Hélène Metzger to Emile Meyerson, 1933', *Isis* 94/3 (2003): 477–91.

11 Her letters to Sarton shed a new light on the lines of *Newton, Stahl, Boerhaave* (Paris Alcan, 1974 [1930]), in which she expressed her gratefulness to Meyerson (in relation to the chapters on Stahl and Boerhaave) and her dedication of the volume to him. In these letters, she ironically commented on the fact that Meyerson thought that the Stahl chapter was the best one because he had supervised it, and when she proposed this chapter for pre-publication in *Isis*, she said that Meyerson thought that it was too condensed, but she in fact condensed it further: see Metzger's letters to Sarton, 10 August 1925, and of 23 September 1926, Houghton Library, bMS Am 1803 (1032).

12 Metzger's letter to Sarton 14 April 27, Houghton Library, bMS Am 1803 (1032).

13 Hélène Metzger, 'La philosophie d'Emile Meyerson et l'histoire des sciences', *Archeion* 11 (1929): xxxii–xliii (xxxii), my emphasis. Léon Brunschvicg was present at her talk, and participated in the discussion that followed (see pp. xlii–xliii).

14 Full French text and English translation of this letter is published in *Isis*: Chimisso and Freudenthal, 'A Mind of Her Own: Hélène Metzger to Emile Meyerson, 1933'. Gad Freudenthal unearthed this letter. For an insight in the biography and personality of Meyerson, see Eva Teklès's research based on the Meyerson archives in Jerusalem: Eva Telkès-Klein, 'Emile Meyerson, d'après sa correspondance. Une première ébauche', *Revue de synthèse* 125, 5th ser. (2004): 197–215; and Eva Telkès-Klein, 'Emile Meyerson: A Great Forgotten Figure', *Iyyun* 52 (2003): 235–44.

Neither a Mere Philosopher's Laboratory nor an Accumulation of Facts: Metzger's History of Scientific Thought

The problems between Metzger and Meyerson probably originated in their incompatible personalities, and the difficult position that she, as a woman, had in an overwhelmingly masculine world. However, there also were more scholarly reasons for Metzger to want to distinguish her work from his. At the Paris conference mentioned above, Metzger said that she could not 'declare herself his disciple', because the problems in which she was most interested were 'not at all those that his epistemological work ha[d] aimed to clarify'. It is not by chance that Metzger stressed that Meyerson's work was 'epistemological': she was presenting not only the difference between her own research and Meyerson's, but also the tensions between the more historically oriented and the more philosophically oriented approaches to the history of science. Indeed, she placed Meyerson in a tradition of philosophers who 'used' the history of science in order to derive information about the mind. This programme, she continued, had been 'systematically' developed in the nineteenth century by Comte, Cournot and Renouvier. Metzger also mentioned Ernst Mach and, 'closer to us', Pierre Duhem, her 'colleague' Abel Rey and Léon Brunschvicg. Metzger's comments shed an interesting light on how these scholars regarded one another's projects. As discussed in the previous chapter, Rey was critical of Brunschvicg's and Meyerson's approaches to science, because he deemed them too philosophical. Similarly, in a rather damning review, Febvre judged Brunschvicg's *L'esprit européen* as the typical work of a historian of philosophy, who has a general and selective approach to history and who ignores 'our' [the historians'] history.[15] He regarded Rey's first volume of *L'encyclopédie française*, which covered similar ground to Brunschvicg's *L'esprit européen*, as quite a different effort, for its insistence on the importance of history in the understanding of the current *outillage mental*, and because it included chapters by a linguist and a mathematician.[16] In fact, for Metzger there was little difference between Rey, Meyerson, Brunschvicg and Duhem. To her, they were all philosophers with an interest in the history of science.

15 Febvre's unwillingness to consider the philosophers' approaches to history has been seen by Roger Chartier as one the gravest limitations of his method. He believes that if Febvre had taken seriously the epistemological model offered by Bachelard, Koyré and Canguilhem, he would have avoided drawing sometimes simplistic conclusions from statistical enquiry: see Roger Chartier, *Cultural History: Between Practices and Representations* (trans. Lydia G. Cochrane; Ithaca: Cornell University Press, 1988), p. 35. Important elements of Bachelard's method were already present in Brunschvicg's work, and Koyré gave talks at the Centre de synthèse, but Febvre did not seem interested in his work, nor in Bachelard's, despite the latter having been a student of his associate Rey.

16 Lucien Febvre, 'Un cours de Léon Brunschvicg' [1947], in Lucien Febvre, *Combats pour l'histoire* (Paris: Colin, 1992), pp. 289–90. For a modern reader, it is difficult to regard Rey's account of intellectual history as more precise and scientific than Brunschvicg's, as Febvre suggested. Just to give an example, Febvre argued that Brunschvicg, who intended to analyse the European mind, had not clearly defined what 'European' meant. It is easy to agree with Febvre, but, on the other hand, it is hard to regard Rey's 'western thought' and 'our thought' as less vague and undefined.

She pointed out that, like the other philosophers, Meyerson aimed to formulate a 'theory of human reason', but his works were not 'directly or exclusively' dedicated to the history of science.[17]

She made it clear that she did not intend to use the history of science simply as a repository of examples to support philosophical theories. The punctiliousness of her textual analysis is only one of the signs of how seriously she took historical accuracy. In her books, there is an abundance of long quotations, made to illustrate her point, or presented to be explained and commented on. French history of chemistry was not unused to commentary of texts.[18] However, Metzger's close analysis of sources is more marked than that of many of her contemporaries, and not only those whom she considered philosophers. Her philological attitude and her penchant for historical accuracy resulted in the narrow focus of her works, which, except for her popular *La Chimie*,[19] examine seventeenth and eighteenth-century chemical and allied disciplines. This is of course very different not only from grand works such as Duhem's *Le Système du Monde* or from Sarton's *Introduction*,[20] and even from Brunschvicg's sweep of the whole of history of mathematics, but also from the work of her colleagues at the Centre de synthèse. Mieli for large part of his life planned to complete a general history of science, and published on the history of chemistry as well as on Arabic science[21] while Abel Rey's interests ranged from ancient science to modern physics and philosophy.

Despite her focus on the history of chemistry, Metzger wholeheartedly shared the Centre de synthèse's ideal of 'historical synthesis'. Indeed, although the growth of her theoretical awareness over the years is evident, already in her first book she argued that the history of crystallography, just like that of any science, is linked to the general history of humanity.[22] Moreover, she took extremely seriously theories that twenty years later Gaston Bachelard would call 'lapsed', or dropped out of

17 Metzger, 'La philosophie d'Emile Meyerson et l'histoire des sciences', p. xxxiii.

18 Marcelin Berthelot, the most important historian of chemistry of the previous generation, had contributed greatly to the publication of primary sources, ranging from Arabic and Syrian alchemic manuscripts to records of Lavoisier's laboratory: Marcelin Berthelot, *Collections des anciens alchimistes grecs* (2 vols; Paris: Steinheil, 1887); Marcelin Berthelot, *La chimie au Moyen âge, vol. I: essai sur la transmission de la science antique au Moyen âge* (Paris: Imprimerie nationale, 1893); Marcelin Berthelot, *La chimie au Moyen âge, vol. II: l'alchimie syriaque* (Paris: Imprimerie nationale, 1893); Marcelin Berthelot, *La chimie au Moyen âge, vol. III: la chimie arabe* (Paris: Imprimerie nationale, 1893); Marcelin Berthelot, *La révolution chimique; Lavoisier* (Paris: Alcan, 1890). See also his works of introduction to ancient chemistry and alchemy: Marcelin Berthelot, *Les origines de l'alchimie* (Paris: Steinheil, 1885); Marcelin Berthelot, *Introduction a l'étude de la chimie des anciens et du Moyen âge* (Paris: Steiheil, 1889). See also Milhaud's use of long quotations in: Gaston Milhaud, *Leçons sur les origines de la science grecque* (Paris: Alcan, 1893).

19 Hélène Metzger, *La Chimie*, (Paris: Boccard, 1930).

20 Pierre Duhem, *Le Système du Monde: Histoire des doctrines cosmologiques de Platon à Copernic* (10 vols; Paris: Hermann, 1913–59); George Sarton, *Introduction to the History of Science* (3 vols; Baltimore: William and Wilkins, 1927–47).

21 Mieli, *La science arabe et son rôle dans l'évolution scientifique mondiale.*

22 Metzger, *La genèse de la science des cristaux*, p. 224.

modern science. All her work was going to be a study of past science from the point of view of the past rather than that of modern science.[23] It was therefore not just to pay tribute to the spirit of Rey's Institut d'Histoire des Sciences et Techniques that she introduced her 1932–33 course at that Institut by stating the following five fundamental points: that chemistry did not develop independently of general history; that chemistry was not an isolated science; that chemistry was not independent of philosophical currents and revolutions; and finally that chemical literature had strong links with general literature. She concluded her preliminary remarks of her lectures at that institution on the relationship between chemical literature and general history with the following words:

> No point of human history is irrelevant for the history of the sciences; social and political history, history of philosophy, history of literature including history of theatre ... history of industry and trade, to which could be also added history of art and history of occultism, are [all] useful to the researcher who proposes to study the chemists' writings.[24]

Although she claimed that the history of chemistry should be linked with all other aspects of history, the reader does not find a presentation of a whole culture in her books. She does indeed refer to general philosophy, or more particular events such as the creation of academies,[25] but in fact her books are very focused on chemical doctrines and practices. Why did she then criticize Mieli for not connecting 'scientific thought' to 'general thought'?[26] I think this is because she believed that by showing the

23 Ilana Löwy has linked Metzger's 'constructivist' approach with her early scientific training in crystallography, and in particular with her study of classifications, and the need to legitimize the by then out-of-date brand of science that she had studied. The positive part of Löwy argument is important and deserving great attention. Bernadette Bensaude-Vincent has similarly argued that Metzger's object of study, the history of chemistry, was crucial in the development of her philosophy, as it was for Duhem, Meyerson and Bachelard (Bernadette Bensaude-Vincent, 'Chemistry in the French Tradition of Philosophy of Science: Duhem, Meyerson, Metzger and Bachelard', *Studies in History and Philosophy of Science* 36 [2005]: 627–48). However, I think that Löwy's suggestion that Metzger wrote her first book with no knowledge of history and philosophy, and that therefore all her ideas must have come from science goes a little too far. Metzger had been married to a historian and we know that her uncle Lévy-Bruhl, who was a historian of philosophy, helped her with her research. I think it is safer to assume Metzger's knowledge of some philosophy and history from which she derived ideas which indeed were very widespread at that time in France, rather than invoking the 'influence' of a 'general relativistic sensibility that prevailed in the 1920s'. See Ilana Löwy, 'Constructivist Epistemologies: Metzger and Fleck', in Freudenthal (ed), *Etudes sur/Studies on Hélène Metzger*, pp. 219–35 (231).

24 Hélène Metzger, 'La littérature chimique française aux 17e et 18e siècles', *Thalès* 2 (1935): 162–5 (163–4). Metzger reworked the lectures on Lavoisier that she delivered at the Institut d'histoire des sciences et techniques in 1932–33 into a small volume in Abel Rey's series 'Exposés d'Histoire et Philosophie des Sciences': Hélène Metzger, *La philosophie de la matière chez Lavoisier* (Paris: Hermann, 1935).

25 See Hélène Metzger, *Les doctrines chimiques en France du début du XVIIe à la fin du XVIIIe siècle* (Paris: Presses Universitaires de France, 1969 [1923]), p. 345.

26 Hélène Metzger, Letter to Sarton of 5 August 1939, Houghton Library, bMS Am 1803 (1032).

mentalities of seventeenth and eighteenth-century chemists, she aimed to disclose a way of thinking of a whole civilization. In her view, this was possible because those chemists inevitably shared their ways of thinking with their contemporaries. She argued that, although past chemical theories at first seem unconnected to each other, their analysis in fact revealed their common characteristics, which allow the historian to define the 'intellectual activity' of a whole age.[27]

She always aimed to describe the general characteristics of a civilization, and this is why her books are always somewhat 'choral'; she consistently refused to write a history of 'great men'. As Bensaude-Vincent has put it with reference to *La Chimie*, 'the only portrait Metzger deigns to provide is a sort of identikit picture of the early seventeenth century chemist'.[28] Her approach was significantly different from those of the director of the Section d'histoire des sciences Aldo Mieli, and of George Sarton. In *Le doctrines chimiques*, she explicitly wrote that she had avoided all biographical information, and comments on the psychology of the individual authors, because the history of chemistry, as she saw it, was independent of the individual scholars' lives;[29] in *Newton, Stahl, Boerhaave*, she followed the same principle (despite the title), but added short biographical notices in the Appendix, following Sarton's suggestion.[30] She had given way to the pressure from the latter, but only in a formal way.

Metzger was not interested either in the history of great men or in the history of events, and she opposed to the latter her 'history of scientific thought'.[31] Her focus on ideas was at odds with the practice of the history of science as accumulation of 'positive' facts. She aimed to capture ideas and doctrines as they were elaborated by the authors she studied, rather than as codified and established doctrines. Although she opposed the use of the history of science as a repository of examples at the service of the philosopher, she equally rejected a merely empirical approach to history; in fact, her ultimate goal was to understand mental processes behind scientific theories and practices. She was intellectually close to other philosophers beside Meyerson, and there was one in particular whom she did not mind calling 'my teacher':[32] her uncle Lucien Lévy-Bruhl, who 'always encouraged' her in her research and who had

---

27 Metzger, *Les doctrines chimiques en France du début du XVIIe à la fin du XVIIIe siècle*, p. 27.

28 Bernadette Bensaude-Vincent, 'Hélène Metzger's *La Chimie*: a Popular Treatise', *History of Science* 25/1 (1987): 71–84 (74).

29 Metzger, *Les doctrines chimiques en France du début du XVIIe à la fin du XVIIIe siècle*, p. 10.

30 Metzger, *Newton, Stahl, Boerhaave*, p. 11.

31 Hélène Metzger, 'La méthode philosophique dans l'histoire des sciences', *Archeion* 19 (1937): 204–16 (205). Metzger's theoretical articles and extracts from her reviews (including those cited in the present chapter) have been reprinted in Hélène Metzger, *La méthode philosophique en histoire des sciences. Textes 1914–1939, réunis par Gad Freudenthal* (Paris: Fayard, 1987). I will refer to the original versions because many of the historiographical articles were papers delivered at the Centre de synthèse and as such were followed by discussions, which were published in *Archeion* alongside the papers. These discussions are very relevant to my argument.

32 Metzger, Letter to Sarton of 18 May 22, Houghton Library, bMS Am 1803 (1032).

always 'accepted to be the first reader' of her manuscripts.[33] Indeed, she did not mind compiling the indexes of his books,[34] just as he did not mind requesting conference registration forms on her behalf.[35]

Lévy-Bruhl's work, as she put it, had been discussed and employed not only by ethnologists, but also by 'philosophers, psychiatrists, psychologists, pedagogues, sociologists and historians'.[36] The historian Metzger was indeed among those who employed Lévy-Bruhl's theory of mentality. Just as Lévy-Bruhl believed that so-called primitive people behaved and reasoned according to categories that were different from those of modern European people, so Metzger thought that her alchemists followed different principles from those of modern science. If she had applied the logic of modern science to them, she would have had to conclude that most of them were lunatics, who not only seemed to follow lines of reasoning that look patently incorrect, but also insisted on pursuing research programmes that trial after trial failed to achieve the desired results. For instance, the 'philosophers of metals' insisted on looking for a solution to the problem of the transmutation of metals, despite the regular and total failures of their experiments.[37] Metzger was interested not only in describing their practices, but also in finding out the worldviews and ways of thinking that led to them. She explained that the alchemists never discussed the possibility of the transmutation of one metal into another (ideally into gold), because they took it for granted. She argued that at the core of the alchemists' belief was the assumption that metals were a 'natural class' and that within this class each element was linked to another by 'analogy'. In fact, their whole worldview was dominated by analogy.

She regarded her chemists' use of analogy as one their most fundamental ways of understanding their objects and the relationship between them. In *Les concepts scientifiques*, she discussed analogy extensively. In this theoretical book, she analysed

---

33 Metzger, Letter to Sarton of 14 April 27, Houghton Library, bMS Am 1803 (1032). It may also be significant that as a young woman she went with the manuscript of her first book to her uncle's colleagues (after failing with her own teachers) rather than for instance her husband's, who would have been historians. In fact, we know that she found her husband's friends' attitude towards her (many years after her husband's death) rather upsetting, as they did not take her scholarly work seriously. She analysed their attitude as class snobbery: in her view they regarded her as a rich man's daughter who had 'bought' herself an 'aristocrat', in the sense of an intellectual. They even insinuated that she had paid Sarton to publish her articles (she had devolved part of her Bourdin prize to *Isis*; however, in other cases, she was paid for her reviews): see Metzger, Letter to Sarton of 22 April 1926 Houghton Library, bMS Am 1803 (1032).

34 See Lévy-Bruhl, *L'âme primitive*, 'Avant-propos'; Lucien Lévy-Bruhl, *Le surnaturel et la nature dans la mentalité primitive* (Paris: Presses Universitaires de France, 1963 [1931]), p. xiii; Lucien Lévy-Bruhl, *L'expérience mystique et les symboles chez les primitifs* (Paris: Alcan, 1938), p. 301.

35 See Lévy-Bruhl's letter dated 21 December 1921, in Merllié, 'Les rapports entre la *Revue de métaphysique* et la *Revue philosophique*', Appendix.

36 Metzger, 'La philosophie de Lévy-Bruhl et l'histoire des sciences': 18.

37 Metzger, *Les doctrines chimiques en France du début du XVIIe à la fin du XVIIIe siècle*, ch. 2.

the ways in which past natural philosophers divided objects into classes and saw the relationships between these objects. She argued that the three main ways in which they organized their data were analogy, permanence of substances, and evolution, which were in turn subdivided in further categories. 'Active analogy', or action of similar on similar, was the subcategory on which she reflected the most, because it was the type of relation that she found most frequently in her seventeenth and eighteenth-century sources. She often identified 'active analogy' with Lévy-Bruhl's 'law of participation', and regarded it as fundamental to what she called 'expansive thought', which was similar, but by no means identical, to Lévy-Bruhl's 'primitive mentality', as I shall discuss below.[38]

The Primitive Core of Modern Science

Although she contrasted her own use of history with the use of it made by philosophers, she not only argued that philosophers and historians need to collaborate,[39] but she also claimed that the objective of history of science was 'to improve the knowledge of the human mind'.[40] Indeed, both her drive to understand her chemists' mental mechanisms, and her aim to generalize her findings were long-lasting. Most of her theoretical papers were first delivered at the Section d'histoire des sciences of the Centre de synthèse, where the audience was only partially receptive. It was there that she presented her general view of the mind, in her talk on the 'mental *a priori*'. Her use of the expression '*a priori*' was rather ambiguous. Indeed, she had to clarify that the meaning provided by Lalande's *Vocabulaire de philosophie*,[41] that is, any item of knowledge not acquired through experience, was more restrictive than her own. Her concept of mental *a priori* includes not only the notions 'preceding experience' but also the 'fundamental tendencies that generate those notions'.[42]

Metzger did not clarify further what these 'fundamental tendencies' were, and indeed it seems that she referred to something that could not be defined precisely. They did include intellectual categories, such as Kant's, as well as non-rational and emotional predispositions that impact upon the way people reason. However, unlike Kant, and in agreement with Lévy-Bruhl and Brunschvicg, she believed that 'the *a priori* is not and cannot be the same in all times and places; ... there is not only one *a priori*, but several *a prioris*, very different from one another and sometimes heterogeneous and incompatible'.[43] One of the reasons why on this occasion she preferred to employ the expression 'mental *a priori*' rather than 'mentality', which she often used as well, is probably that she wanted to indicate a deeper mental level

38 Metzger, *Les concepts scientifiques*, pp. 37–8. See also her review of *L'âme primitive*, in which she wrote that 'it was Lévy-Bruhl's works that inspired' her to elaborate her theory of active analogy: Hélène Metzger, 'Lucien Lévy-Bruhl, *L'âme primitive*', *Isis* 9 (1927): 482–6 (486, n. 1).

39 Metzger, 'La philosophie d'Emile Meyerson et l'histoire des sciences': xlii.

40 Metzger, 'La méthode philosophique dans l'histoire des sciences': 207.

41 Lalande, *Vocabulaire technique et critique de la philosophie*.

42 Hélène Metzger, 'L'*a priori* dans la doctrine scientifique et l'histoire des sciences', *Archeion* 18 (1936): 29–42 (33).

43 Metzger, 'L'*a priori* dans la doctrine scientifique et l'histoire des sciences', p. 33.

than that of mentality. In other words, she referred to human 'tendencies' that would generate observable ways of thinking, or mentalities.[44] Just as Metzger's 'mental *a priori*' is close but not identical to Lévy-Bruhl's 'mentality', her 'expansive thought' is close but not identical to his 'primitive mentality'. Indeed, she regarded their two conceptions as two interpretations of the same phenomenon. She said that she did not want to go as far as Lévy-Bruhl and claim that 'expansive thought' was prelogical and mystical. Rather, for her what marked 'expansive thought' was its lack of reflectivity, or self-criticism.[45]

She did not see a complete opposition between primitive and modern scientific thought,[46] but rather regarded 'expansive' (rather than 'primitive') thought as the locus of creativity, and as providing the 'the first inspiration' of 'the most beautiful discoveries' and 'most admirable inventions'.[47] In her lectures at the Ecole pratique des hautes études, she argued that 'spontaneous or expansive thought'[48] was at the roots of Newton's theory of universal attraction. She explained that Newton based his achievements not only 'on Kepler's astronomy' and 'on Hooke's coherent hypotheses', but also 'on Henry More's metaphysics, on the Neoplatonic mystics' beliefs [and] on the astrologers' dreams'.[49] Indeed, through universal attraction, God, who had previously been excluded from the functioning of nature, re-entered it. Newton, she explained, superimposed on Boyle's corpuscular philosophy a force that bodies exert on one another without contact. He did not deduce this force from *a priori* principles, but rather assumed its existence. For Metzger this idea of action of one thing on another without contact did not come from scientific rationality, but rather from the mind's 'instinctive activity', that is from what she called 'expansive thought'; in her eyes, the proof of this was that we find the same type of idea throughout history. The type of reasoning behind the elaboration of this force was for her the same as that observed by Lévy-Bruhl in 'primitive' societies, and could also be found in Renaissance therapeutics, which admitted the action of similar on similar; it was finally not unlike the Neoplatonic doctrines that assumed that

44 On Metzger's mental *a priori*, see Gad Freudenthal, 'Hélène Metzger (1888–1944)', in Michel Bitbol and Jean Gayon (eds), *L'épistémologie française, 1830–1970* (Paris: Presses Universitaires de France, 2006), pp. 113–56. For a critical view of Metzger's use of 'mental *a prioris*' and her search for fundamental mental 'tendencies', see Golinski, 'Hélène Metzger and the Interpretation of Seventeenth Century Chemistry'.

45 Metzger, 'L'*a priori* dans la doctrine scientifique et l'histoire des sciences', p. 37.

46 Metzger, 'La philosophie de Lévy-Bruhl et l'histoire des sciences', p. 19.

47 Ibid., p. 21.

48 Hélène Metzger, *Attraction universelle et religion naturelle chez quelques commentateurs anglais de Newton. Première partie. Introduction philosophique* (Paris: Hermann, 1938), p. 9. Metzger used the expressions 'expansive thought' and 'spontaneous thought' as synonyms.

49 Ibid., p. 8. She also published her other lectures: Hélène Metzger, *Attraction universelle et religion naturelle chez quelques commentateurs anglais de Newton. Deuxième partie. Newton – Bentley – Whiston – Toland* (Paris: Hermann, 1938); Hélène Metzger, *Attraction universelle et religion naturelle chez quelques commentateurs anglais de Newton. Troisième partie. Clarke – Cheyne – Derham – Baxter – Priestly* (Paris: Hermann, 1938). The Ecole pratique des hautes études awarded her a diploma based on these books.

aqua regia had a 'friendship' with gold. She regarded universal attraction simply as another instance of what she called 'active analogy'.

However, the difference between Newton and the astrologers was that he 'chained' this force to a series of rigorous calculations. For Metzger, although at the core of a scientific doctrine there may be, and often is, an 'inspiration' that comes from the same type of approach to reality that Lévy-Bruhl described as 'primitive', this inspiration must be evaluated critically and rationally, and successfully included in a rational and mathematical framework. Reflective thought, as described by Metzger, verifies its knowledge and itself, is 'constantly polemical' and says 'no' to whatever is obscure and unbelievable. In this respect, 'reflective thought' is opposed to 'expansive thought'; nevertheless, Metzger thought that the former extended the 'élan' provided by the latter.[50]

Mental a prioris *and the Historian's Mentality*

Although she believed that 'expansive thought' was universal and timeless, Metzger regarded past doctrines as the product of historically specific mentalities, or, in her terminology, 'mental *aprioris*'. These mental *aprioris* could for her be 'very different from one another' and even 'heterogeneous and incompatible'.[51] Metzger applied the self-critical attitude of science to her own work as a historian, and interrogated herself on how she could bridge the gap between her own way of thinking and that of the authors whom she studied. In her own words, she wondered whether the historian of science should make himself a 'contemporary of the scientists of whom he speaks'.[52] That her answer was in the affirmative, there is no doubt. However, she described the 'disorientation' that the historian felt when reading old texts, the difficulty of identifying the very objects under discussion, and the uselessness of applying 'positive logic' to them, for, if read according to modern logic, they would look like the product of 'folly'.[53] The importance that she accorded not only to logic, but also to emotions and values in the formation of science led her to argue that the historian should try to recreate the mental states [*états de l'âme*] of past scientists, as well as their 'doubts, disappointments and triumphs'.[54]

---

50 Metzger, 'L'*a priori* dans la doctrine scientifique et l'histoire des sciences', p. 38. As I shall discuss in the next chapter, Gaston Bachelard also characterized scientific reason as 'polemical reason', and as saying 'no' to previous experience. However, unlike Bachelard, Metzger did not regard expansive thought as an obstacle to science. The terminology that she employed to describe the work of 'reflective thought', characteristic of science, is close to the terminology that Gaston Bachelard used to elaborate his more complex doctrine of the scientific mind.

51 Ibid., p. 33.

52 I use the masculine pronoun in order to translate Metzger's sentence literally. Cf. the title of one of her talks at the Centre de synthèse: Hélène Metzger, 'L'historien des sciences doit-il se faire le contemporain des savants dont il parle?', *Archeion* 15 (1933): 34–44.

53 Metzger, 'L'*a priori* dans la doctrine scientifique et l'histoire des sciences', pp. 30–31.

54 Ibid., p. 32.

The manner in which she intended to bridge the gap between mentalities raised many concerns in the more positivistic-oriented members of the Centre de synthèse, who sought a fully objective historiography. Her whole notion of mental *a priori* was viewed suspiciously by some, including Henri Berr. '*A priori*' was not an expression that would have pleased Berr, committed as he was to an empiricist view of the historian's craft. He conceded that historians should 'sympathize' with the scientist whom they study, but he did not see why this should imply the *a priori* in the historian. He continued by inviting Metzger to distinguish between two *a prioris*: the scientist's hypotheses, and the 'more or less founded' hypotheses of previous stages of knowledge. Obviously Metzger declined to agree, by saying that Berr's proposal was incompatible with her own doctrine. Berr and Metzger appeared to aim to answer different philosophical questions. Berr must have seen in Metzger's *a priori* either a type of radical rationalism, opposed to his own empiricism, or even conventionalism. Years earlier, Berr's close associate Abel Rey had written an article that sheds light on the lack of philosophical engagement between Berr and Metzger. Rey, discussing the tension between *apriorism* and empiricism, remarked that recently *a priori* had become synonym with arbitrary, and many thought that *a priori* and deductive science is arbitrary and conventional, whereas experimental science is objective.[55]

As far as the work of the historian is concerned, whereas a modern critic such as Gad Freudenthal interprets Metzger's notion of *a priori* as close to Hans Georg Gadamer's 'prejudice', many in her audience must have seen it as a prejudice in the Enlightenment negative meaning of the word.[56] The entry on 'History' that Berr and Febvre jointly wrote for the *Encyclopaedia of the Social Science* provides a good example of the approach that was common and encouraged at the Centre de synthèse. In this entry, Berr and Febvre acknowledged that historians' own ideas, and their historical, social and economic positions, could have an 'influence' on their work, but simply as a 'serious cause of error'. The only remedy, in their view, consisted in the 'progressive broadening of history' and in the achievement of

55 Abel Rey, 'L'*a priori* et l'expérience dans les méthodes scientifiques', *Revue de métaphysique et de morale* 16 (1908): 883–8 (884).

56 Gad Freudenthal, 'Epistémologie des sciences de la nature et herméneutique de l'histoire des sciences selon Hélène Metzger', in Freudenthal (ed.), *Etudes sur/Studies on Hélène Metzger*, pp. 161–88. 'Monsieur Serrus' (almost certainly Charles Serrus, teacher at the *lycée* Louis-le-Grand and later *chargé de conférences* at the Sorbonne, see André Lalande, 'La Philosophie en France: 1945–1946', *The Philosophical Review* 56/1 [1947]: 1–18 [10]), who attended her talk, praised her and agreed that prejudice is a necessary element of invention; however, he soon added that prejudice is also the 'illusion' that science has the right and duty to correct. His argument became closer to Metzger's when he denounced that 'Kantian illusion' of the fixity of knowledge. Section d'Histoire des Sciences, 'Séance du 20 novembre 1935', *Archeion* 18 (1936): 75–9 (76–7). Maxime Laignel-Lavastine (Professor of History of Medicine at the Faculty of Medicine) expressed the opinion that a psychology or maybe psychoanalysis of the *a priori* was needed. Gaston Bachelard was shortly to publish *La Formation de l'esprit scientifique. Psychanalyse de la connaissance objective*, in which he attempted precisely that.

'universal history'.[57] And indeed, Metzger's view of the history of science as part of universal history was welcome, and both Berr and Rey said so; the latter added that he had 'fought all [his] life for the success of the same thesis'.[58]

However, the fact that she did not think that universal history alone could neutralize the effects on history writing of the historians' culture, way of reasoning and values,[59] in short of their 'mental *a prioris*', created a disagreement between herself and the historians, first of all Lucien Febvre, with whom she could have had much in common. Indeed, Metzger's fundamental questions seem to be rather close to those of Lucien Febvre and Marc Bloch.[60] For instance, just as she aimed to find the way of reasoning that lead the alchemist to insists on their quest despite their lack of success, so in the *Les rois thaumaturges* Bloch aimed to understand what led the people of France and England to continue believing for centuries in the therapeutic virtues of the royal touch.[61] Febvre in *Le problème de l'incroyance au XVI*^e^ *siècle, la religion de Rabelais* aimed to demonstrate the impossibility for Rabelais, as for any of his contemporaries, of conceiving the world without God; similarly, Metzger argued that the alchemists could not think of a natural world in which all things were not in a relation of mutual correspondence and influence. And yet, Febvre hardly ever ventured to the historians of science's meetings at the Centre. He did attend one of Metzger's talks, on whether 'the different aspects of the same period of a civilization (literature, sciences, arts)' could be considered as 'projections' of the same 'mental state'. In his only minuted intervention, he dismissed Metzger by saying that the problem that she had posed was for him already solved, as he thought synchronically.[62]

---

57 Henri Berr and Lucien Febvre, 'History', in Edwin R.A. Seligman (ed.), *Encyclopaedia of the Social Sciences*, vol. 7 (New York: Macmillan, 1932), pp. 357–68 (367).

58 Section d'histoire des sciences, 'Séance du 21 avril 1937', *Archeion* 19 (1937): 254–8 (255). I have briefly compared Metzger's and Rey's view of mentality and history in Cristina Chimisso, 'Hélène Metzger: The History of Science between the Study of Mentalities and Total History', *Studies in History and Philosophy of Science* 32/2 (2001): 203–41.

59 After the discussion of her *La méthode philosophique dans l'histoire des sciences*, Metzger received a letter from one of those who had attended, Bénézé, who regarded his conception of value as part of what Metzger called *a priori*, and she responded by expressing her agreement, see Section d'histoire des sciences, 'Séance du 21 avril 1937', pp. 257–8, n. 1.

60 For an evaluation of the difference between Febvre's and Bloch's conceptions of history and mentalities, see André Burguière, 'La notion de mentalité chez Marc Bloch and Lucien Febvre: deux conceptions, deux filiations', *Revue de synthèse* 111 (1983): 333–48.

61 Marc Bloch, *Les rois thaumaturges: étude sur le caractère surnaturel attribué à la puissance royale, particulièrement en France et en Angleterre* (Strasbourg: Faculté des Lettres de l'Université de Strasbourg, 1924).

62 Section d'histoire des sciences, 'Séance du 19 mars 1930', *Archeion* 12 (1930): 375–79 (379). Incidentally, the world of the *Annales*, which Febvre had just founded with Marc Bloch, was not more friendly to women than French universities were: Natalie Zemon Davis, while stressing the (often unacknowledged) work that wives and female students carried out as research assistants, notices that only two women published in the journal between its foundation in 1929 and Bloch's death in 1944: Natalie Zemon Davis, 'Women and the World of the *Annales*', *History Workshop Journal* 33 (1992): 121–37 (122).

In her talks at the Centre de synthèse, she intended to further the discussion on the variety of mentalities, and on the manner in which historians should read old texts, not only considering their distance in time from their sources, but also the particularity of their own mentalities. The Section d'histoire des sciences of the Centre turned out not to be the best place to do so. Both the director of the unit Aldo Mieli, and the director of the Centre itself, Henri Berr, did not regard favourably what they probably saw as Metzger's relativism and idealism. Indeed, she wrote to Sarton that she had been told that she was too 'abstract' and that she replaced history with metaphysics.[63]

Metzger thought that old texts were the product of different ways of reasoning and different worldviews. She believed that the aim of her work was to uncover how differently the minds of the alchemists worked from those of the modern historian, who in her view was not a neutral observer, but was rather immersed in a particular mentality. Metzger saw her own work as a part of a wider history of the mind, and as a contribution to the task of improving our knowledge of the mind. She aimed to answer similar questions to those of Lévy-Bruhl and Brunschvicg. She amply acknowledged her debt to Lévy-Bruhl's philosophy, but made clear her departure from his view of mentalities. She often referred to Brunschvicg, and wrote reviews of his books, but she engaged less with his philosophy than with that of Lévy-Bruhl.[64] In comparison with Lévy-Bruhl and Brunschvicg, Metzger considerably lessened the contrast between modern and scientific mentality, on the one hand, and religious and 'primitive' mentality, on the other. Lévy-Bruhl's representation of modern mentality, implicit by contrast with primitive mentality, is scientifically oriented and thoroughly secular. Brunschvicg's distinct take on the history of the mind is similar in that he regarded the modern way of thinking as secular, and indeed for him progress meant a progressive advancement of secularism. Metzger shared the two philosophers' secular beliefs. However, in her view, the distinction between intuitive and religious ways of thinking, on the one hand, and scientific thought, on the other, was more blurred. She did not regard science as completely opposed to mystical, religious or 'primitive' thought, but rather saw the latter as a possible

63 Metzger, Letter to Sarton, 1 November 1937, Houghton Library, bMS Am 1803 (1032). Since it seems that this comment had been made about the book on Lavoisier that Metzger had just written, it was perhaps Mieli who objected to her method, as he himself had written on Lavoisier; Metzger, *La philosophie de la matière chez Lavoisier*; Aldo Mieli, *Lavoisier* (Genova: A.F. Formiggini, 1916).

64 Hélène Metzger, 'Léon Brunschvicg, *L'expérience humaine et la causalité physique*', *Isis* 5 (1923): 479–83; Hélène Metzger, 'Léon Brunschvicg, *De la connaissance de soi*', *Archeion* 13 (1931): 389–91; Hélène Metzger, 'Léon Brunschvicg, *La physique du vingtième siècle et la philosophie: le rôle du pythagorisme dans l'évolution des idées*', *Archeion* 19 (1937): 301–303; Hélène Metzger, 'Léon Brunschvicg, *Les âges de l'intelligence*', *Archeion* 16 (1934): 254–5 (extracts of the first three reviews are published in Metzger, *La méthode philosophique en histoire des sciences. Textes 1914–1939, réunis par Gad Freudenthal*). At least on one occasion, Brunschvicg sent one of his books to Metzger: see Metzger, Letter to Sarton, 14 April 1927, Houghton Library bMS Am 1803 (1032). Metzger attended Brunschvicg's lectures at the Sorbonne, and was a friend of his wife, see Freudenthal, 'Hélène Metzger: Eléments de biographie', p. 199.

source of inspiration for science. She still made a clear distinction between the two, as she attributed to scientific thought the self-reflectivity, logical procedures and the need to verify assertions that 'expansive thought' lacked. However, Metzger detected God at the core of Newton's theory, and did not see why religious, mystical and non-rational ideas in general could not be at the origin of science itself. In Metzger's view of science there was not the militant secularism that characterized Lévy-Bruhl's, Brunschvicg's and Rey's doctrines. Alexandre Koyré went even further in erasing the line between science on the one hand and metaphysics and religion on the other.

Alexandre Koyré and the History of Intellectual Revolutions

A Cosmopolitan Historian of Philosophical, Religious and Scientific Ideas

When, in January 1935, Hélène Metzger proposed Alexandre Koyré as a new member of the Section d'histoire des sciences of the Centre de synthèse,[65] he was 42 years old and Director of Studies in History of Religious Ideas in Modern Europe at the Ecole pratique des hautes études, 5th *section* (Religious Sciences).[66] He had published books on the idea of God in the philosophies of Descartes[67] and of Anselm,[68] on the philosophy of the medieval mystic Jacob Boehme,[69] and on philosophy and the national problem in Russia at the beginning of the nineteenth century.[70] He had also translated Anselm's *Fides quaerens intellectum* and published articles on Hegel.[71] His post and his publishing record may look rather at odds with the remit of the Section d'histoire des sciences. However, in the previous few years, Koyré

65 Section d'histoire des sciences, 'Séance du 23 Janvier 1935', *Archeion* 17 (1935): 81–4 (82). Koyré was duly elected.

66 A clear and informative presentation of the Ecole pratique des hautes études was offered by Koyré himself, see Alexandre Koyré, 'L'ecole pratique des hautes études', in Koyré, *De la mystique à la science*, pp. 6–17.

67 Alexandre Koyré, *Essai sur l'idée de Dieu et les preuves de son existence chez Descartes* (Paris: Leroux, 1922). This was the thesis of his diploma in religious sciences at the Ecole pratique des hautes études.

68 Alexandre Koyré, *L'idée de Dieu dans la philosophie de saint Anselme* (Paris: Leroux, 1923). This was the thesis of his *doctorat d'Université.*

69 Alexandre Koyré, *La philosophie de Jacob Boehme. Etudes sur les origines de la métaphysique allemande* (Paris: Vrin, 1929). This was the thesis of his *doctorat d'Etat.*

70 Alexandre Koyré, *La philosophie et le problème national en Russie au début di XIXe siècle* (Paris: Honoré Champion, 1929). This was the complementary thesis of his *doctorat d'Etat.*

71 Saint Anselm, *Fides quaerens intellectum. Introduction, texte, traduction par A. Koyré* (Paris: Vrin, 1927); Alexandre Koyré, 'Rapport sur l'état des études hégéliennes en France', *Revue d'histoire de la philosophie* 5/2 (1931): 147–71; Alexandre Koyré, 'Note sur la langue et la terminologie hégéliennes', *Revue philosophique* 56 (1931): 409–39; Alexandre Koyré, 'Hegel à Jena', *Revue philosophique* 59 (1934): 274–83, all repr. in Alexandre Koyré, *Etudes d'histoire de la pensée philosophique* (Paris: Gallimard, 1971).

had shifted his interest to Copernicus and Galileo. Indeed, he was then working on what was going to become arguably his most famous book: the *Etudes galiléenes*. Metzger was already aware of Koyré's value, as she had been attending his lectures on Galileo at the Ecole pratique des hautes études since 1934; she continued until 1939, first as a 'regular auditor' and then as 'graduate student'; she also took 'active part' in Koyré's course entitled 'Galilean studies'.[72] In 1937–38, Koyré spent most of the academic year at the University of Cairo, and arranged to be substituted in his lectures by Metzger and Alexandre Kojève.

The central place that Koyré came to occupy within the history of science might suggest that any talk of 'marginality' should be out of place. However, as Georges Canguilhem pointed out, in France Koyré's scholarly career developed in an 'intellectual milieu' which was at the periphery of 'the [French] traditional university milieu'.[73] His *curriculum studiorum* and his social position in France differed from the golden standards of the most important French academics. His marginality was more marked than that of Lévy-Bruhl and Brunschvicg. Jewish like Lévy-Bruhl and Brunschvicg, although, unlike them, from a well-to-do family,[74] he was a foreigner, notwithstanding his French citizenship acquired sometime after the First World War, during which he had served in the French army and subsequently in the Russian army.[75] His education differed considerably from that of his French counterparts. In particular, before gaining his French formal qualifications, from his *diplôme d'études supérieures* to his doctorate, and studying with the most important French philosophers, including Bergson, Brunschvicg, Lalande and Delbos, Koyré spent important years in Germany, where he studied with Husserl and Hilbert.[76] There, from 1910, he joined the so-called 'Göttingen group', a discussion group formed around Husserl, which employed phenomenology as a tool to investigate a broad range of issues.[77]

---

72 Koyré, *De la mystique à la science*, pp. 43ff.

73 Georges Canguilhem, 'Preface', *History and Technology* 4 (1987): 7–10 (7).

74 See Pietro Redondi and P.V. Pillai (eds), *The History of Science: The French Debate* (London: Sangam Books, 1989), p. 255. See also Bernard I. Cohen and René Taton, 'Hommage à Alexandre Koyré', in Bernard I. Cohen and René Taton (eds), *Mélanges Alexandre Koyré, I: L'aventure de la science* (Paris: Hermann, 1964), pp. xix–xxv.

75 It is not clear when Koyré acquired his French citizenship; it is believed to be just after the war, but reportedly there are no archival sources to prove it; in a *curriculum vitae* dated 1940, he declared he was a French citizen, see Redondi in Koyré, *De la mystique à la science*, p. 5.

76 Ibid.; Suzanne Delorme, 'Hommage à Alexandre Koyré', *Revue d'histoire des sciences* 18/2 (1965): 129–39 (129–30).

77 See Gérard Jorland, *La science dans la philosophie: les recherches épistémologiques d'Alexandre Koyré* (Paris: Gallimard, 1981), pp. 27–42; and Ethan Kleinberg, *Generation Existential: Heidegger's Philosophy in France 1927–1961* (Ithaca and London: Cornell University Press, 2005), pp. 27–8; Paola Zambelli, 'Alexandre Koyré alla scuola di Husserl a Gottinga', *Giornale Critico Della Filosofia Italiana* 78/3 (1999): 303–54; Karl Schuhmann, 'Alexandre Koyré et les phénoménologues allemands', *History and Technology* 4 (1987): 149–65.

Koyré's knowledge and first-hand experience of the most recent developments of German philosophy was certainly superior to that of the French historians of philosophy. Moreover, as I have discussed, the latter were keen to oppose their approaches to German idealism. Koyré himself, in a 1930 conference, contrasted the lack of a French neo-Hegelian school with the development of neo-Hegelian philosophies in Germany, England and Italy. The neglect of Hegel for him led to an early return to Kant, hence the strength of neo-Kantian philosophies in France.[78] Koyré rightly pointed out that the prestige of scientific thought in France was at odds with 'Hegel's anti-mathematism'; similarly, his philosophy of history was not compatible with the French historical school. He recalled Brunschvicg's disparaging judgement of Hegel as the 'master of contemporary scholastic' and his dismissal of the 'absurd disdain of German romanticism for scientific knowledge'.[79] He did not forget the philosophers who had shown interest in Hegel, including Boutroux, Lévy-Bruhl and Delbos, but at the same time noticed that the First World War interrupted their interest, which was often replaced by open hostility towards German culture.[80] Koyré's familiarity not only with Hegel's philosophy,[81] but also with the German philosophy of his time, made him a 'bridge between Germany and France, Husserl and Bergson'.[82] Indeed, the journal *Recherches philosophiques*, which he co-edited, seemed to play this role: Alexandre Kojève and Jean Wahl reviewed books on phenomenology, there was a section dedicated to 'existence and situation' and existentialist philosophers published in it, including Jean-Paul Sartre and Gabriel Marcel. The journal's innovative outlook was by no means aimed to challenge the French philosophical establishment, who in fact supported it: its advisory board included all the major historians of philosophy: Bréhier, Brunschvicg, Gilson, Lévy-Bruhl, Rivaud and Robin, and also Bouglé, Lalande and Rey. Bachelard, who started to co-edit the journal in 1934, reviewed books on logic and epistemology, and published articles in it from the first volume.[83]

---

78 Koyré, 'Rapport sur l'état des études hégéliennes en France', p. 147–9. The situation of Hegelian studies in France greatly changed in the 1930s, as Koyré acknowledged in a postscript to this article published in Koyré, *Etudes d'histoire de la pensée philosophique*, pp. 249–50. See also Roberto Salvadori, *Hegel in Francia: Filosofia e politica nella cultura francese del Novecento* (Bari: De Donato, 1974); M. Roth, *Knowing and History: Appropriation of Hegel in Twentieth-Century France* (Ithaca and London: Cornell University Press, 1988); and Michael Kelly, *Hegel in France* (Birmingham: Birmingham Modern Languages Publications, 1992).

79 Koyré, *Etudes d'histoire de la pensée philosophique*, pp. 227–8.

80 Ibid., pp. 230–36.

81 In addition to the already cited articles on Hegel, see Alexandre Koyré, 'Hegel en Russie', *Monde Slave* 1 (1936): 397–88. For a full bibliography of Koyré's works, see Jean-François Stoffel, *Bibliographie d'Alexandre Koyré* (Firenze: Olschki, 2000).

82 Kleinberg, *Generation Existential*, p. 59.

83 Gaston Bachelard, 'Idéalisme discursif', *Recherches philosophiques* 4 (1934–35): 21–9; Gaston Bachelard, 'Le monde comme caprice et miniature', *Recherches philosophiques* 3 (1933–34): 306–20; Gaston Bachelard, 'Logique et épistémologie', *Recherches philosophiques* 6 (1936–37): 410–13; Gaston Bachelard, 'Noumène et microphysique', *Recherches philosophiques* 1 (1931–32): 55–65.

Koyré did not use his knowledge of modern German philosophy in order to develop his own philosophy along Hegelian, phenomenological or existentialist lines. His approach to Hegel was that of a historian of philosophy: he was interested in studying, understanding and presenting Hegel's thought, rather than appropriating it for his own purposes. In particular, he never subscribed to a Hegelian philosophy of history.[84] Moreover, his interpretation of Hegel has nothing of the existentialist and romantic leanings of Jean Wahl's reading, which was crucial in the so-called Hegel Renaissance of the 1930s.[85] Despite his role as editor of *Recherches philosophiques*, Koyré's work has little to do with the new philosophies, due to such thinkers as Jean-Paul Sartre, Maurice Merleau-Ponty and Emmanuel Levinas, which developed in France as a result of the cross-fertilization of French philosophy with the philosophies of Hegel, Husserl and Heidegger.[86] Unlike those French philosophers, Koyré was a historian of ideas and approached modern German philosophy in the same way as he approached any other text.

Textual Analysis and the History of Intellectual Revolutions

Koyré's 'attentive reading of texts' has been admired and praised,[87] although his exegeses have been carefully checked, and there have been many claims of errors

84 See Pietro Redondi, 'Henri Berr, Hélène Metzger et Alexandre Koyré: la religion d'Henri Berr', in Biard, Bourel and Brian (eds), *Henri Berr et la culture du XXe siècle*, pp. 139–57 (148).

85 See Koyré's interpretation of Hegel's early writing in Koyré, 'Hegel à Jena' and Wahl's in Wahl, *Le malheur de la conscience dans la philosophie de Hegel*. For a comparison, see Roth, *Knowing and History*, pp. 5ff.

86 Koyré also wrote on Heidegger: see his 'L'évolution philosophique de Martin Heidegger' [1946], in Koyré, *Etudes d'histoire de la pensée philosophique*, pp. 271–304. See also Alexandre Koyré, 'Introduction [of M. Heidegger "Qu'est-ce que la métaphysique"]', *Bifor* 8 [1931]: 5–8). Writing to the translator, Henry Corbin, he declared that he had found Heidegger's text 'incomprehensible' and 'boring': Alexandre Koyré, '[Letter to Henry Corbin, 1937 or 1938]', in Christian Jambet (ed.), *Henry Corbin* (Paris: L'Herne, 1981), p. 330. Later, he criticized both Heidegger's language and Corbin's translation of the key-term 'Dasein' as 'realité humaine', see Koyré, 'L'évolution philosophique de Martin Heidegger' [1946] in Koyré, *Etudes d'histoire de la pensée philosophique*, p. 272. On Koyré's view of Heidegger, see Paola Zambelli, 'Alexandre Koyré, "Present Trends of French Philosophical Thought": Introduction', *Journal of the History of Ideas* 59/3 (1998): 521–30. Zambelli writes that Koyré invited Heidegger to give a talk in France in 1929, but he did not go. When he did give a talk in France, in 1955, Koyré did not attend. Koyré 'condemned and made known Heidegger's Nazism outside France', as early as 1933, in Levinas's testimony: Zambelli, 'Alexandre Koyré, "Present Trends of French Philosophical Thought": Introduction', pp. 525–6.

87 Yvon Belaval, 'Les recherches philosophiques d'Alexandre Koyré', *Critique* 20/206 (1964): 675–704 (675); Jorland, *La science dans la philosophie: les recherches épistémologiques d'Alexandre Koyré*, pp. 78–90; Jacques Le Goff, 'Histoire des sciences et histoire des mentalités', *Revue de synthèse* 104 (1983): 405–15 (412); Nicholas Jardine, 'Koyré's Intellectual Revolution', *La lettre de la Maison française d'Oxford* 13 (2001): 11–25. Jardine has also argued that Koyré's 'meticulous *explication des textes*' was combined

and biases.[88] Whatever the view of his philological accuracy, it is without doubt that the '*explication des textes*' was at the core of Koyré's method. His dedication to textual analysis distinguished him from many philosophers, but was consistent with that of the historians of philosophy. Like Brunschvicg's, his approach to the history of science had its roots in the history of philosophy, with its consequent attention to the interpretation of texts, their terminology and above all the ideas expressed in them. As a historian of philosophy, he had translated and edited philosophical works. His linguistic competence, which included Russian, German, French, Italian, English and Latin, has often been noted.[89] He shared with Metzger his method based on textual analysis, but this was only one of the similarities that she must have seen between his works and her own.

Metzger's philological attitude made her dialogue with the historians of the Centre de synthèse difficult, especially because they regarded her approach as limited to ideas and texts; despite her methodological claims, in practice she neglected social, economic and material circumstances. On the other hand, her philosophical approach, and her preoccupation with avoiding anachronism, were met with almost incomprehension by Mieli and Berr, as discussed above. Koyré was both an interpreter of texts and a philosopher, and she believed that his aims as a historian of science were similar to hers. Already in his work on Galileo, which was being published in the *Annales de l'Université de Paris*, Koyré had made clear

---

with 'bold, often speculative, reconstruction on the basis of scattered clues': Nicholas Jardine, 'Koyré's Kepler/Kepler's Koyré', *History of Science* 38 (2000): 363–77 (364).

88 Among those who have criticized his exegeses, see M. Finocchiaro, 'Logic and Scholarship in Koyré's historiography', *Physis* 19 (1977): 5–27; and Maria Rosa Antognazza, 'Leibniz and the Post-Copernican Universe. Koyré Revisited', *Studies in History and Philosophy of Science* 34 (2003): 309–27. Thomas Kuhn named Koyré as one of the sources of inspiration for his *Structure of Scientific Revolutions* (Thomas S. Kuhn, *The Structure of Scientific Revolutions* [Chicago: The University of Chicago Press, 1996 (1962)]), but he pointed out Koyré's mistakes and biases, and claimed that Koyré, from his sick bed had indicated one book (which obviously is *The Structure*) as opening the future of the history of science: Thomas S. Kuhn, 'Alexandre Koyré and the History of Science', *Encounter* 34 (1970): 67–9. On Kuhn's claims, see Jardine, 'Koyré's Intellectual Revolution', pp. 363–4; on the links between Koyré's and Kuhn's historiographies, see Brendan Larvor, 'Why did Kuhn's *Structure of Scientific Revolutions* Cause a Fuss?', *Studies in History and Philosophy of Science* 34 (2003): 369–90; and Stefan Amsterdamski, 'La philosophie en science', *History and Technology* 4 (1987): 103–13.

89 See for instance Belaval, 'Les recherches philosophiques d'Alexandre Koyré', p. 875. Koyré's linguistic competence was also appreciated during his life, and he was often requested to act as a translator. Lévy-Bruhl successfully asked him to translate the Schwarztkoppen papers, that is, the papers of the German military attaché in Paris, which confirmed Alfred Dreyfus's innocence: Maximilian von Schwartzkoppen, *Les carnets de Schwartzkoppen: la vérité su Dreyfus, édité par Bernhard Schwartfeger et traduits sur le texte allemand par Alexandre Koyré; préface Lucien Lévy-Bruhl* (Paris: Rieder, 1930). Bachelard asked Koyré to translate Heidegger's *Being and Time*, or to find a translator, but he realized that it would not be easy to find the 'heroic translator' needed for such task. See Bachelard's letter of 4 June 1938 in Christian Jambet (ed.), *Henry Corbin* (Paris: Editions de l'Herne, 1981), p. 311. Koyré indeed did not translate it.

his 'philosophical' interest in the history of science, and cited Duhem, Meyerson, Cassirer and Brunschvicg to validate it. He argued that:

> The study of the evolution (and the revolutions) of scientific ideas ... shows us the human mind at grips with reality, reveals to us its defeats and victories; shows us what superhuman effort each step on the way to knowledge of reality has cost, effort which has sometimes led to a veritable 'mutation' in human intellect, that is to a transformation as a result of which ideas which were 'invented' with such effort by the greatest of minds become accessible and even simple, seemingly obvious, to every schoolboy.[90]

Koyré acknowledged his debt to Gaston Bachelard for the concept of mutation of human intellect.[91] Like Bachelard, Brunschvicg and Metzger, Koyré aimed to show, through the study of the history of science, the changes in the way human beings reason. In other words, as for them, for him the mind has a history. This obviously does not mean that his solutions overlapped with theirs, which in turn were rather different from one another. In particular, Koyré was rather careful to avoid relativism, and indeed even appealed to the authority of Meyerson, the champion of the fixity of the mind, when he argued that it is by studying what 'shocks' us in other ways of thinking that we understand what is permanent in the mind.[92] He however dedicated a large part of his work to explaining the changes in ways of seeing the world and thinking. In particular, his preoccupation with explaining how it came to pass that ideas that did not exist at all in a certain period, later became obvious to all, must have convinced Metzger that their objectives were very similar. In an article on Galileo that he published a few years after that meeting, he set the aim of his study as follows:

> We have ... to understand why, and how, the principle of inertial motion, which to us appears so simple, so clear, so plausible and even self-evident, acquired this status of self-evidence and *a priori* truth whereas for the Greeks as well as for the thinkers of the Middle

90 This passage was first published in Alexandre Koyré, 'Au l'aurore de la science moderne. La jeunesse de Galilée (I)', *Annales de l'Université de Paris* 10 (1935): 540–41 (540); this article was going to become the Introduction to the *Etudes galiléenes* (Alexandre Koyré, *Etudes galiléennes* [Paris: Hermann, 1939]). The English translation (Alexandre Koyré, *Galileo Studies* [trans. John Mepham; Hassocks: The Harvest Press, 1978], p. 1) renders 'humain intellect' in this quotation as 'human thought'. I prefer the literal 'intellect': his reference to Bachelard (see below in the main text) suggests that Koyré saw a change in mental faculties, rather than simply in the exercise of these faculties. Lalande's *Vocabulaire* defines 'intellect' as 'higher cognitive faculty', opposed both to sensation and intuition, and indicates 'understanding' as the English translation of this term (Lalande, *Vocabulaire technique et critique de la philosophie*, vol. 1, p. 521). I chose 'intellect' rather than 'understanding', because it is how '*intellect*', notably in Bergson's work, is normally translated.

91 In the *Annales* article he cited Gaston Bachelard, *Le nouvel esprit scientifique* (Paris: Presses universitaires de France, 1991 [1934]); in his 1939 volume, he added Gaston Bachelard, *La formation de l'esprit scientifique: contribution à une psychanalyse de la connaissance objective* (Paris: Vrin, 1993 [1938]).

92 A. Koyré in Section d'histoire des sciences, 'Séance du 23 Janvier 1935', p. 83. For a comparison of Meyerson's and Koyré's views, see Mario Biagioli, 'Meyerson, Science and the Irrational', *Studies in History and Philosophy of Science* 19 (1988): 5–42.

Ages the idea that a body once put in motion will continue to move forever, appeared as obviously and evidently false, and even absurd.[93]

The aim that Koyré expressed here was the mirror-image of Metzger's objective of understanding how certain concepts, like 'active analogy', or procedures, like the transformation of one metal into another, were tacitly accepted for a long time, but then became simply absurd to everyone. Despite their different perspectives, their aims were very close indeed.

Koyré's analysis of conceptual changes also recalls that of Léon Brunschvicg. Just as Brunschvicg saw a sharp difference between Newton's and Einstein's respective conceptions of space, so Koyré analysed the previous historical mutation, from the Aristotelian conception to that of classic physics. He argued that after Galileo, space was no longer a concrete space, in which objects occupy their given place, but rather the abstract space of Euclidean geometry. Koyré would dedicate most of his work to the history of the intellectual revolution that lead to modern science. He regarded this revolution as a slow process, which for him began as early as the fourteenth century, when the early Italian humanists, such as Petrarch, started to display a lack of interest in Aristotelian scholastics.[94] The Renaissance marked a further step, as for him it prepared Galileo's and Descartes' revolution; this in turn lead to Newton's physics. Koyré represented this revolution first of all as an intellectual change, which brought about a mutation in aims, values and worldview. As he argued in the English-language work *From the Closed World to the Infinite Universe*, the world itself changed. The Aristotelian and medieval world had the earth at its centre, enclosed in concentric spheres, on which the planets were carried. This world was ontologically differentiated: the heavenly spheres were incorruptible and moved circularly and eternally, whereas the sub-lunar world was corruptible, and its motions were rectilinear. In Koyré's account, the scientific revolution substituted this model with an open and infinite universe, in which there is no ontological difference between celestial and terrestrial bodies, and consequently the same physics applies to heaven and earth. In this new infinite universe, the earth lost its central place.[95]

For Koyré, the scientific revolution had not only implemented a dramatic change of worldviews, but also a change in questions and methods. The post-Galilean world is for Koyré a mathematical world; consequently the aim of post-Galilean science is to measure, rather than to establish qualitative differences among things. Indeed, for Koyré the language used to investigate nature changed with the scientific revolution: it became a mathematical language, which served to study this new universe made

93 Koyré, 'Galileo and Plato' [1943], repr. in Alexandre Koyré, *Metaphysics and Measurement; Essays in Scientific Revolution* (Cambridge, MA: Harvard University Press, 1968), p. 19.

94 Koyré, 'La pensée moderne' [1930], in Alexandre Koyré, *Etudes d'histoire de la pensée scientifique* (Paris: Presses Universitaires de France, 1966), p. 19.

95 On Koyré's notion of scientific revolution, see Gérard Jorland, 'La notion de révolution scientifique: la modèle de Kuhn', in Michel Bitbol and Jean Gayon (eds), *L'épistémologie française, 1830–1970* (Paris: Presses Universitaires de France, 2006): 157–71.

of quantities rather qualities.[96] The revolution that Koyré narrated was for him in the minds of sixteenth and seventeenth-century Europeans; in his view, it 'changed the very framework and patterns of our thinking'.[97] Koyré's interest in collective representations, which was already present in his dissertation on Anselm,[98] was close to that of the historians of the mind, and in particular to Metzger's. Like her, he appeared to be writing a chapter in the history of the mind through the history of scientific thought.

Metzger, like her fellow members of the Centre de synthèse, firmly believed that the history of science could not be separated from all other aspects of history. Koyré, who worked on the history of religious ideas, the history of philosophy and the history of science, was showing by historical examples that she and her colleagues were right. In his *curriculum vitae*, written in 1951 for his candidature to the Collège de France, Koyré declared that 'from the beginning' of his research he had been 'inspired' by his 'conviction of the unity of human thought' and that he had always thought it impossible to separate the history of philosophical thought from religious thought, as well as from scientific thought. Indeed, he argued that Boehme's mysticism would be completely incomprehensible without Copernicus's new cosmology.[99] It is clear that he saw no discontinuity between his work on Boehme and his work on Copernicus, which he had already started at the time of his joining the Centre de synthèse.[100] He also presented himself as the inheritor of the French philosophical tradition of studying scientific thought and its history, after the example of 'Tannery, Duhem, Hannequin and Brunschvicg, Meyerson and Pierre Boutroux'.[101]

I have no doubt that Metzger hoped that Koyré could be an ally within the Section d'histoire des science, and could help her to shift the balance of power.[102] Another historian of the mind, and a brilliant one at that, could have lent authority to her

96 See the collections of articles later published as Koyré, *Metaphysics and Measurement*.

97 Alexandre Koyré, *From the Closed World to the Infinite Universe* (New York: Harper, 1958 [1957]), p. v. Although writing in the 1950s, Koyré was referring to his studies thus far.

98 See Paola Zambelli, 'Alexandre Koyré versus Lucien Lévy-Bruhl: From Collective Representations to Paradigms of Scientific Thought', *Science in Context* 8/3 (1995): 531–55. Zambelli has also noticed that Koyré cited Lévy-Bruhl in his dissertation. The title of Zambelli's article may be puzzling, for the article is primarily aimed at showing Lévy-Bruhl's influence on Koyré, rather than their opposition. This article is a translation from Italian (Paola Zambelli, 'Alexandre Koyré e Lucien Lévy-Bruhl. Dalle rappresentazioni collettive ai paradigmi del pensiero scientifico', *Intersezioni* 13 [1993]: 395–409). The title of the original Italian article reads 'Alexandre Koyré *and* Lucien Lévy-Bruhl', rather than '*versus*'.

99 Koyré, 'Orientation des recherches et projets d'enseignement' [1951], in Koyré, *De la mystique à la science*, p. 127; a long extract has also been published as 'Orientation et projets des recherches' in Koyré, *Etudes d'histoire de la pensée scientifique*.

100 Alexandre Koyré, 'Copernic', *Revue philosophique* 58 (1933): 101–18.

101 Koyré, *De la mystique à la science*, p. 131. This passage is missing in Koyré, *Etudes d'histoire de la pensée scientifique*.

102 Zambelli has argued that Metzger's methodology can offer the key to Koyré's: Zambelli, 'Alexandre Koyré versus Lucien Lévy-Bruhl', p. 544.

position. When the Section d'histoire des sciences discussed Koyré's talk on the young Galileo, she said that his talk 'marvellously illustrated' what she had argued a month earlier, that is, that the experimental method at the beginning was a rejection of common experience and hearsay.[103] Mieli, who chaired the session, sat in silence reserving his quite harsh criticism of the 'newcomer' for an article on Galileo that he published two years later. Mieli argued that Koyré's interpretation of Galileo's work was 'simplistic and often wrong'; he was particularly incensed by Koyré's claim that Galileo had not performed his famous experiments, including that at the leaning tower of Pisa.[104] If Metzger did have a strategy to steer the research programmes of the Section, her move to attract Koyré to it did not work. The Section itself did not survive for long, although the Comité d'histoire des sciences did, as Mieli directed it from Argentina. However, it is hard to see it as a coincidence that Koyré was only elected to it a month after Mieli's death, in 1950.[105] In 1955, at Pierre Sergescu's death, Koyré was elected director (*secrétaire perpétuel*).[106]

In the years between his talk at the Centre de synthèse and the war, Koyré continued his teaching at the Ecole pratique des hautes études, and then in Cairo, where he wrote part of his *Etudes galiléennes*.[107] In the following years, international events had a dramatic impact on the group. In 1939 Mieli fled to Argentina. Writing to him in 1942, Sarton claimed that he had tried unsuccessfully to arrange a visa for

103 Section d'histoire des sciences, 'Séance du 22 janvier 1936', *Archeion* 18 (1936): 238–47 (239).

104 Aldo Mieli, 'Il tricentenario dei "Discorsi" di Galileo Galilei', *Archeion* 21 (1938): 193–297 (249–50, n. 4). (This footnote, like the title, is in Italian, while the main text of the article is in French). Mieli was referring to Alexandre Koyré, 'Galilée et l'expérience de Pise: à propos d'une légende', *Annales de l'Université de Paris* 12 (1937): 422–53, now reprinted in Koyré, *Etudes d'histoire de la pensée scientifique*. Not much can be read in Mieli's lack of participation in Koyré's session, for only the following year he resigned as director of the Section, on the basis that his deafness prevented him from chairing the sessions in a satisfactory way, see Section d'histoire des sciences, 'Communications officielles', *Archeion* 19 (1937): 402.

105 See Académie Internationale d'Histoire des Sciences, 'Elections de 1950', *Archives Internationales d'Histoire des Sciences* 12 (1950): 657.

106 Académie Internationale d'Histoire des Sciences, 'Election du Secrétaire perpétuel', *Archives Internationales d'Histoire des Sciences*, 33 (1955): 436. Apparently Bodenheimer had proposed to Taton that he should become secretary, but he refused, and the latter was instead appointed associate secretary: see Friederich Simon Bodenheimer, Letters to Sarton of 13 March 1955 and 23 June 1955, Houghton Library, bMS Am 1803 (159). Taton thought that Bodenheimer should have replaced Sergescu, see René Taton, Letter to Sarton of 11 January 1955, Houghton Library, bMS Am 1803 (1655); see also Delorme, 'Hommage à Alexandre Koyré', p. 138.

107 In a letter from Cairo, Koyré wrote that 'thanks to Lalande' he had more time to write his work on Galileo. See Koyré, Letter to Corbin, in Koyré, '[Letter to Henry Corbin,1937 or 1938]'; in 1939, when Lalande was invited to lecture at the University of Cairo, he sent Koyré to replace him, for he did not want to leave France with the war looming: Delorme, 'Hommage à Alexandre Koyré', p. 135.

Metzger.[108] She stayed in France, and was killed by the Nazis. Koyré, on the other hand, was invited by the Italian Jew Max Ascoli, Professor at the New School of Social Research, to join the faculty there.[109] Sarton wrote a letter of reference for him.[110] In New York, Koyré not only taught at the New School, but also took part in the foundation of Ecole libre des hautes études, along with other Francophone scholars in exile.[111] The link with the United Stated continued after the war was over. Koyré did indeed resume his old job in Paris, but continued delivering lectures in the United States, in several universities.[112] From 1955, he divided his time between Paris and Princeton. Koyré enjoyed an international career and fame, to the point where he came to be seen as the contender for the title of founder of the history of science as a discipline (along with Sarton), in particular in the United States.[113] His role in French history of science is less clear: on the one hand, the importance of his work is undisputed, as is his aim to implement a renewal of the discipline; on the other hand, his failure to secure a chair at the Collège de France is a reminder of his ambiguous place in the French academic milieu.[114]

Between History and Philosophy: Koyré's Ambiguous Place in French Post-War Academia

In France, the intellectual milieu in which the projects of writing the history of the mind developed had changed by the time Koyré went back to Paris after the war. Metzger had died, and so had Brunschvicg and Abel Rey. At any rate, it would be wrong to think that Koyré, despite his successes, could inherit the leadership of this type of studies, which would have required a powerful academic position, of the type that Brunschvicg and Bréhier had held at the Sorbonne, and Bergson and Gilson at the Collège de France. He did submit his candidature to the Collège de France for a chair of History of Scientific Thought, as replacement of Gilson's chair, but, despite Lucien Febvre's support, the Collège de France preferred Martial Gueroult.[115] The scholars whom he listed in his letter of candidature to the Collège de France, as

108 Georges Sarton, Letter to Mieli, 10 August 1942, Houghton Library, bMS Am 1803.1 (413).

109 Koyré, *De la mystique à la science*, p. 61.

110 An extract of Sarton's letter is published in ibid., p. 63.

111 Paola Zambelli has provided an insightful account of Koyré's role in the Ecole libre d'hautes études, Paola Zambelli, 'Filosofia e politica nell'esilio: Alexandre Koyré, Jacques Maritain e l'Ecole Libre a New York', *Giornale Critico Della Filosofia Italiana* 18/1 (1998): 73–112.

112 For a personal account of Koyré's activities in the USA, see Bernard I. Cohen, 'Alexandre Koyré in America: Some Personal Reminiscences', *History and Technology* 4 (1987): 55–70, and for more general biographical information, Cohen and Taton, 'Hommage à Alexandre Koyré'.

113 Jardine, 'Koyré's Intellectual Revolution', p. 20.

114 See Redondi in Koyré, *De la mystique à la science*, pp. ix–xxvii.

115 For Febvre's letter and an extract of Koyré's letter of application, see ibid., pp. 125–34.

forming the tradition in which he placed his own work is revealing.[116] None of these scholars were alive after the war, and the only two who could have been relevant in terms of enduring influence on higher-education institutions were Léon Brunschvicg and Pierre Boutroux. The latter had held the chair of History of the Sciences at the Collège de France, and therefore it was particularly important for Koyré to present himself as his heir. However, this was really a little far-fetched, for Pierre Boutroux, as discussed in Chapter 4, was the champion of 'pure' history of mathematics, unencumbered by philosophy. On the other hand, Brunschvicg's base of power was the Sorbonne, and his legacy there had been assured before his death when, in 1940, his student Gaston Bachelard was appointed to a chair.

René Taton has judged as 'ambiguous' Koyré's relationship with the Sorbonne Institut d'histoire des sciences and techniques,[117] which was the place for philosophers who studied the history of science. These were the scholars who continued that tradition of which Koyré claimed to be the heir in his letter of candidature to the Collège de France. Yet, his links to that institution appeared to have been rather lukewarm. Koyré did not publish in *Thalès*, the journal of the Institut, and, although he was a member of the Institut's management committee, so were about fifty other Parisian professors, as Taton has remarked.[118] His courses at the Ecole pratique des hautes études were part of the curriculum of the students of the Institut, but nevertheless Koyré did not play an important role in the activities of the Institut either before or after the war. After the war, if anything, his links to the Sorbonne appeared to grow weaker: whereas in his 1951 application to the Collège de France, Koyré claimed that the history of scientific thought was one of the 'most precious' traditions of French philosophy, in 1957, in his presentation of his project for a Centre for the history of science and technology at the Ecole pratique des hautes études, he argued that France was 'deplorably behind' in the study of the history of the sciences compared with the United States, England, Belgium and Holland, 'despite the existence – at least on paper – of an Institute of History of Science and Technology at the University of Paris'.[119]

It was not only a problem of alliances and patronage. Koyré's brand of the history of the mind occupied a difficult place epistemologically, almost a no-man's land. The study of the history of the mind was indeed at the crossroads of sociology, history, philosophy and the history of science. Nevertheless, within this interdisciplinary field, different approaches had taken shape. Koyré seemed to share aims with all of them, but in fact to be comfortable in none. It is well known that Koyré refused strict sociological explanations of scientific change. He did interpret scientific theories as expressions of ways of thinking that neither belong only to the individual scientists,

---

116 As mentioned above, these were Tannery, Duhem, Hannequin, Brunschvicg, Meyerson and Pierre Boutroux.

117 René Taton, 'Alexandre Koyré et l'essor de l'histoire des sciences en France (1933 à 1964)', *History and Technology* 4 (1987): 37–53 (40). Taton has also judged as difficult Koyré's relationship with the Section d'Histoire des Sciences at the Centre de synthèse, and the Comité international d'Histoire des Sciences. This is not surprising considering his difficulties with Aldo Mieli, who headed both institutions.

118 Ibid., p. 40.

119 Koyré, *De la mystique à la science*, pp. 135–6.

nor only to scientists, but rather to a larger community. Because of this approach, Yehuda Elkana has regarded him as posing the basis for the historical sociology of scientific knowledge.[120] However, while conceding that some social conditions are necessary to the development of science, above all the existence of a leisured class with the time to pursue knowledge for its own sake, Koyré rejected any causal explanation based on social structures. In his own words:

> It ... seems to me vain to attempt to deduce the existence of Greek science from the social structure of the city state, or even from the agora. Athens does not explain Eudoxus, or Plato, any more than Syracuse explains Archimedes; or Florence, Galileo. I even believe, indeed, that the same is true for modern times, and even of the present century despite the so much closer co-operation between pure and applied science ... The social structure of England in the seventeenth century cannot explain Newton, any more that the Russia of Nicholas I can throw light on the work of Lobachevsky, or the Germany of Wilhelm II enables us to understand Einstein. To look for explanations along these lines is an entirely futile enterprise, as futile as trying to predict the future evolution of science or of the sciences as a function of the structure of their social contexts.[121]

He carefully considered, but ultimately rejected, Pierre-Maxime Schuhl's 'psycho-sociological' explanation of the stagnation of technology in the ancient world, because social structures, even though indirectly, still played an important role.[122] Koyré's own account of why ancient and medieval societies did not develop technology in the modern sense of it, relies on his description of the evolution of mental attitudes. For him, the development of technology requires 'the application of the rigid, exact and precise notions of mathematics' to reality. Prior to the scientific revolution this notion was considered absurd. The world of every-day life is not a world of exact measurements, and ancient and medieval people for Koyré only saw the (sub-lunar) world in every-day terms. He argued that human beings lived in a different world from the world that modern science has created; they lived in the world of 'more or less', in which precision was not important. Koyré commented that in the pre-scientific world, it would have looked absurd to employ precise measurements: in his example, a horse is definitely bigger than a dog, but neither of them have 'strictly determined dimensions'.[123] For him, the difference between

---

120 Yehuda Elkana, 'Alexandre Koyré: Between the History of Ideas and Sociology of Disembodied Knowledge', *History and Technology* 4/1–4 (1987): 115–48.

121 Alexandre Koyré, 'Commentary', in A.C. Crombie (ed.), *Scientific Change: Historical Studies in the Intellectual, Social and Technical Conditions for Scientific Discoveries and Technical Invention, from Antiquity to the Present* (London: Heinemann, 1963), pp. 847–57 (855–6).

122 See Pierre-Maxime Schuhl, *Machinisme et philosophie* (Paris: Presses Universitaires de France, 1969 [1938]). Pierre-Maxime Schuhl (1902–1984), *maître de conférences* at Montpellier, then professor at Toulouse and finally at the Sorbonne, was the co-editor of the *Revue philosophique* from 1952, and its sole editor from 1956.

123 Alexandre Koyré, 'Du monde de l'"à-peu-près" à l'univers de la précision' [1948], in Koyré, *Etudes d'histoire de la pensée philosophique*, p. 342.

ancient and medieval people on the one hand, and modern people on the other, is one of 'mentality'.[124]

In Koyré's account, social structures and material tools play secondary roles. He was not satisfied with Lucien Febvre's isolation of obstacles that Renaissance people had to face in order to develop science and technology rather than crafts. Febvre argued that sixteenth-century men had 'no weapons, no tools, no general plan' for 'conquering the world's secrets'. In particular, they lacked precision tools, like the telescope that would enable Galileo to observe the stars, the thermometer, precise clocks and a mathematical language which could be applied to physics.[125] For Koyré, by contrast, it was not the thermometer that was missing, but the idea that heat can be measured with precision. Indeed, Koyré claimed that instruments are the 'embodiment of the mind' and the 'materialisation of thought'.[126] It is then not entirely surprising that Febvre responded to Koyré's remarks by saying that his argument leant towards 'absolute idealism'.[127] For Febvre, the historian should indeed investigate what is possible or impossible for a member of a certain civilization to think, but also give due consideration of the availability of material instruments. For Koyré ideas and theories have precedence on technological inventions and crafts.

Although, after the war, Koyré appeared to be closer to historians than philosophers, his focus on ideas and texts, and ultimately his disregard for social, economic and political events, made his method similar to that of the historians of philosophy. Indeed, in the eyes of many historians, Koyré remained a philosopher, and only a philosopher. Jacques Le Goff, while recognizing Koyré's 'attentive reading of texts', nevertheless argued that the 'great Koyré … could not push his investigation further than the analysis of concepts and systems'. He concluded that:

> He always remained a philosopher, a type of scholar with whom, at least in France, historians have always had difficulty in establishing a dialogue, due to their distrust for the philosophy of history.[128]

Although Le Goff did not go into details, it is not too arduous to guess some of the reasons for his criticism. For a historian of mentalities like himself, the challenge was to examine the links between ideas and social realities, while Koyré seemed to dispense with the latter. To Le Goff, Koyré's analysis of 'concepts and systems' appeared the ultimate objective of his research, while the historian's objective should have been to analyse the historical formation of those concepts and systems, and their possibility of existence and development in a given time and society. Moreover, although Koyré investigated ways of thinking in a manner that was similar to Febvre and was also reminiscent of Lévy-Bruhl, he analysed high culture, and the links that

---

124 Koyré, 'Du monde de l'"à-peu-près" à l'univers de la précision' [1948], p. 348.

125 Lucien Febvre, *The Problem of Unbelief in the Sixteenth Century: The Religion of Rabelais* (trans. Beatrice Gottlieb; Cambridge, MA: Harvard University Press, 1982 [1942]), ch. 11.

126 Koyré, 'Du monde de l'"à-peu-près" à l'univers de la précision' [1948], p. 352.

127 Lucien Febvre, 'De l'à peu près à la précision en passant par l'ouï-dire', *Annales E.S.C.* 5 (1950): 25–31 (26).

128 Le Goff, 'Histoire des sciences et histoire des mentalités', p. 412.

he established were between philosophy, 'science' and religion, the latter conceived as theology rather than popular beliefs. As he himself put it, he believed in the unity of human thought, but he added that this unity applies 'particularly' to thought 'in its highest forms'.[129] By contrast, Febvre's *outillage mental* was aimed precisely at the overcoming of the idea that high culture was independent of the broader culture, in the ethnological sense of the word. Moreover, although Koyré did not put forward, nor believe in, any historical laws, to the historians his presentation of development of ideas, which appeared almost independent of other historical events, must have looked like an idealist construction of history rather divorced from the type of historical accounts that they aimed to provide through empirical research.

On the other hand, Koyré could no longer been seen as an ally by the historians of philosophy who continued their traditional exegesis and explanation of texts. Indeed, the choice of the Collège de France to give the chair left vacant by Gilson to Gueroult rather than to Koyré shows that he could not be seen as a 'mainstream' historian of philosophy. Indeed, Redondi has pointed out that Gueroult was the inheritor of Gilson's idea of *philosophia perennis*, and regarded science as 'a body of truths established outside history'.[130] Koyré's history of science as the history of intellectual revolutions was not compatible with the history of philosophy as practiced by Gilson and his disciples, who continued the tradition that I have analysed in the first two chapters of present book.

It might sound paradoxical, but Koyré ended up being too much of a philosopher for French historians and too much of a historian for French philosophers, in particular for those philosophers who in principle should have been closest to him, that is to say those who continued the tradition of the study of the mind. The equilibrium between the study of the mind that employed the history of science as a laboratory, and the study of the history of science centred on the history of the mind, was precarious. Koyré was a historian of philosophy and religious ideas by training, but for him the history of science was never a laboratory to visit in order to provide an empirical content to his theories about the mind or the development of history. It is not by chance that he did not write extensive theoretical or methodological works, although he did have strongly held ideas about the mind, science and history. His primary goal was that of giving a new and in his view faithful description of the scientific revolution, based on primary sources. By contrast, not only the dominant historians of the mind before the war, above all Lévy-Bruhl and Brunschvicg, but also those who set the intellectual agenda after the war, Gaston Bachelard and Georges Canguilhem, did regard intellectual history as a laboratory for philosophy.

Other, perhaps more subtle, but I think fundamental, differences also separated Koyré from the philosophers who developed the programme of the '*a posteriori*' study of the mind, and who were the constant reference for the practitioners of the philosophical history of science, first of all Lévy-Bruhl and Brunschvicg. Koyré's 'unity of human thought', which at first sight made him close to them, in fact marked a departure from their conceptions of the history of the mind, their philosophy of

129 Alexandre Koyré, 'Orientation des recherches et projets d'enseignement' [1951], in Koyré, *De la mystique à la science*, p. 127; and Koyré, *Etudes d'histoire de la pensée scientifique*, p. 1.

130 Redondi in Koyré, *De la mystique à la science*, pp. xxvi–xxvii, 119.

history and their politics. For Lévy-Bruhl human beings became modern when they embraced scientific rationality, and abandoned magic, superstition and religion. Similarly, Brunschvicg regarded the history of mind as the progressive emergence of rationality, which replaces magic and religious explanations of phenomena. Koyré did not share these philosophers' thoroughly secular vision of modernity and progress; in fact, he regarded religious and scientific ideas as having a shared history, and saw religious ideas at the core of scientific theories. Their political and ethical battles belonged to the Third Republic, and Lévy-Bruhl, Brunschvicg and Rey were all dead before the end of the war. However, the programme of writing the history of the mind had its future assured, due to the appointment of Brunschvicg and Rey's student Gaston Bachelard to the Sorbonne chair of History and Philosophy of the Sciences.

Chapter 6

From the Laboratory to the Tribunal: Historical Epistemology

Gaston Bachelard and the History of the Scientific Mind

Beyond Brunschvicg's Legacy

When Gaston Bachelard obtained the chair of History and Philosophy of the Sciences at the Sorbonne, the directorship of the Institut d'histoire des sciences et techniques, and the editorship of *Thalès*, the space for general history of science had already been secured.[1] It was no longer a matter of establishing a new discipline, nor of justifying a new approach to one's colleagues. Bachelard, and Canguilhem after him, were the heirs of earlier struggles, both epistemological and institutional. Bachelard, however, was not an 'inheritor', as conceived by Bourdieu and Passeron,[2] and indeed his unusual background has been often noticed.[3] He was not only a provincial, but

1 When Bachelard took over as Director, the Institut was not properly functioning, but that was due to the sudden death of Rey, and to the war. For instance, the secretary was a prisoner of war; see Gaston Bachelard, 'Institut d'histoire des sciences et techniques: rapport sur l'année scolaire 1940–41', *Annales de l'Université de Paris* 16 (1941): 325.

2 Pierre Bourdieu and Jean-Claude Passeron, *The Inheritors: French Students and their Relation to Culture* (trans. Richard Nice; Chicago: University of Chicago Press, 1979 [1964]); Pierre Bourdieu and Jean-Claude Passeron, *Reproduction in Education, Society and Culture* (trans. Richard Nice; London: Sage, 1977 [1970]). See Chapter 1 above for a short discussion of the applicability of their model to the first half of the twentieth century.

3 See, for instance, François Dagognet, *Gaston Bachelard: sa vie, son œuvre* (Paris: Presses Universitaires de France, 1965), pp. 1–3; C. Backès-Clément and B. Pingaud, 'Le dernier alchimiste', *L'Arc* 42 (1970): 1–3 (1); J. Brosse, 'Gaston Bachelard. Portrait', *L'Arc* 20 (1962): 92–6 (92). As for all aspects of Bachelard's life and work, the references would be far too many to be listed here. Unfortunately, it is impossible to cite even all the most important works on this philosopher, who has played an important role not only in the history and philosophy of science, but also in literary studies, education and other fields. Here I shall only discuss him as a historian of the mind. For works up to the mid-1990s, the most comprehensive bibliography is H. Choe, *Gaston Bachelard. Epistemologie. Bibliographie* (Frankfurt am Main: Lang, 1994). For an overview of the geographical breadth of Bachelard's reception, see Jean Gayon and Jean-Jacques Wunenburger (eds), *Bachelard dans le monde* (Paris: Presses Universitaires de France, 2000). Among books that have appeared in English, Lecourt, *Marxism and Epistemology. Bachelard, Canguilhem and Foucault*, translated from the original French, had a tremendous impact in the English-speaking world, and presented Bachelard in the Marxist and Althusserian context. Mary Tiles has made Bachelard familiar and clear to Anglo-American philosophers of science: Mary Tiles, *Bachelard: Science and*

he also came from a rural area, and he did not go to Paris to attend any prestigious *lycée*, nor the Ecole normale supérieure, but rather to work for the Post Office.[4] Bachelard's social origin was not completely different from Brunschvicg's or Rey's, as his family had a newsagent and tobacco shop. However, Bachelard's curriculum of studies was more unconventional than theirs. He studied while working at the Post Office, and gained a degree in mathematical sciences. Like Metzger, however, his interests shifted, and while teaching science in a secondary school, he gained a degree in philosophy.[5] He obtained his doctorate at the age of 43, with Abel Rey and Léon Brunschvicg as supervisors of his theses. His choice of professors was excellent: Rey had established general history of science as a discipline with its own institute, journal, *certificat* and diploma, and Brunschvicg, along with Bréhier, was the philosopher whose students enjoyed the most successful careers.[6] Bachelard's theses were soon published by Vrin[7] and reviewed in the *Revue philosophique* by Brunschvicg, who consecrated Bachelard as a philosopher by writing that his methodological assurance, precision and impartiality 'reveal … a thinker of the first order'.[8]

In the 1930s, when Bachelard was Professor of Philosophy at the University of Dijon, as Rey had been in the 1910s, one would nevertheless encounter him in Paris, at the Société française de philosophie[9] as well as the Centre de synthèse. Bachelard

---

Objectivity (Cambridge: Cambridge University Press, 1984). Mary McAllester has discussed all aspects of Bachelard's work: Mary McAllester Jones, *Gaston Bachelard, Subversive Humanist: Texts and Readings* (Madison, Wis.: University of Wisconsin Press, 1991); I have discussed Bachelard's philosophy in the cultural context of the time, and proposed an interpretation of Bachelard's philosophy based on his pedagogical values: Chimisso, *Gaston Bachelard: Critic of Science and the Imagination*.

4 Overviews of Bachelard's life can be found in several publications, see for instance: Georges Davy, 'Gaston Bachelard: l'unité de l'homme et de l'œuvre', *Etudes philosophiques* 2 (1952): 123–33; Dagognet, *Gaston Bachelard: sa vie, son œuvre*; Paul Ginestier, *Pour connaître Bachelard* (Paris: Bordas, 1987); André Parinaud, *Gaston Bachelard* (Paris: Flammarion, 1996).

5 Similarly to Metzger, he lost his spouse very early: he married just before being mobilized to fight in the First World War; his wife died in 1920. She left him a daughter, Suzanne, who would become an important philosopher of science in her own right.

6 See Chapter 1, where I report Bourdieu's findings according to which the most prestigious careers in the 1970s depended on being linked to professors who had been students of either Bréhier or Brunschvicg.

7 Gaston Bachelard, *Essai sur la connaissance approchée* (Paris: Vrin, 1987 [1927]), dedicated to Rey; and Gaston Bachelard, *Etude sur l'évolution d'un problème de physique: la propagation thermique dans les solides* (Paris: Vrin, 1973 [1927]), dedicated to Brunschvicg.

8 Léon Brunschvicg, '*Etude sur l'évolution d'un problème de physique. La propagation thermique dans les solides*, par Gaston Bachelard', *Revue philosophique* 54 (1929): 92–4 (94). The other review was published in the same issue: Léon Brunschvicg, '*Essai sur la connaissance approchée*, par Gaston Bachelard', *Revue philosophique* 54 (1929): 95–101.

9 Bachelard was invited by Brunschvicg to give a talk at the Société: Gaston Bachelard et al., 'Sur la continuité et la multiplicité temporelles. Séance du 13 mars 1937', *Bulletin de la Société française de Philosophie* (1937): 53–81.

was a member of the Centre's unit for 'general synthesis', directed by Rey, and of the Comité international d'histoire des sciences.[10] His membership shows his position in the Parisian intellectual milieu and academic network, but does not mean that he shared either the positivistic views of Berr, or indeed Rey, nor their aim to write a comprehensive and general history.

From the 1930s onwards, his use of history as the laboratory of the philosopher who studies the mind, already present in his early works, became more and more explicit. The titles of the books that he published between 1934 and 1940 do not leave any doubt about his project, as one promised to be about 'the new scientific mind', another about 'the formation of the scientific mind', and the subtitle of a third, published in 1940, presented it as an 'essay of a philosophy of the new scientific mind'.[11] In methodological terms, Bachelard's choice was firmly within the tradition that his teacher Brunschvicg had greatly helped to shape. Indeed the continuity between his own philosophy and that of Brunschvicg has often, and rightly, been highlighted: Wahl described Bachelard as 'the free continuator of Brunschvicg's philosophy' and Gutting has argued that Bachelard's 'view of the relation of science and philosophy derives most directly from Brunschvicg'.[12]

Bachelard himself acknowledged his debt to Brunschvicg, including on a crucial issue, that of the plasticity of the mind. He opposed the 'static conception' of the mind, proposed by Emile Meyerson, to Brunschvicg's dynamic conception.[13] Indeed, he wondered how it was possible that philosophers could still talk about 'fixed frameworks of intelligence and reason', after Brunschvicg's account of the progress and transformation of the mind, to be found in *Etapes de la philosophie mathématique*, *L'expérience humaine et la causalité physique* and *Les âges de l'intelligence*.[14] Like Brunschvicg, Bachelard believed that reason and its object were relative to each other, and indeed transformed each other. However, Bachelard went

10 Bachelard was also a member of one of the committees (on 'issues to be resolved') of the Comité, see Comité International d'Histoire des Sciences, 'Séance de lundí, 27 septembre 1937', *Archeion* 19 (1937): 382–8 (385).

11 Bachelard, *Le nouvel esprit scientifique*; Bachelard, *La formation de l'esprit scientifique*; Gaston Bachelard, *La philosophie du non. Essai d'une philosophie du nouvel esprit scientifique* (Paris: Presses Universitaires de France, 1988 [1940]).

12 Jean Wahl, *Tableau de la philosophie française* (Paris: Gallimard, 1962), p. 114; Gutting, *French Philosophy in the Twentieth Century*, pp. 85–6. See also François Dagognet, 'M. Brunschvicg et Bachelard', *Revue de métaphysique et de morale* 70 (1965): 43–54; Gary Gutting, 'Introduction: What is Continental Philosophy of Science?', in Gary Gutting (ed.), *Continental Philosophy of Science* (Oxford: Blackwell, 2005), pp. 1–16 (4); Jacques Gagey, *Gaston Bachelard ou la conversion a l'imaginaire* (Paris: Rivière, 1969), pp. 30, 54; Carlo Vinti, *Il soggetto qualunque: Gaston Bachelard fenomenologo della soggettività epistemica* (Napoli: Edizioni scientifiche italiane, 1997), pp. 168, 427–52; Teresa Castelão-Lawless, 'Gaston Bachelard et le milieu scientifique et intellectuel français', in Pascal Nouvel (ed.), *Actualité et postérités de Gaston Bachelard* (Paris: Presses Universitaires de France, 1997), pp. 101–15.

13 Gaston Bachelard, *Le rationalisme appliqué* (Paris: Presses Universitaires de France, 1986 [1949]), p. 9.

14 'La philosophie de Léon Brunschvicg' (1949), in Gaston Bachelard, *L'engagement rationaliste* (Paris: Presses Universitaires de France, 1972), pp. 175–6. Bachelard mentioned

further than his teacher, as he himself programmatically declared, in overcoming the distinction between the subject and object of knowledge.[15] Brunschvicg's solution was an idealistic one, and appeared to absorb the object into the subject. Bachelard's own solution has been judged idealistic by some and materialist by others.[16] Some of Bachelard's claims do indeed sound idealistic, for instance when he wrote that something exists by virtue of being (mathematically) thought of, in his words '*cogitatur ergo est*', and when he called sub-atomic particles noumena, and even 'bibliomena', claiming that to exist in books is a type of existence, and a very human existence indeed.[17] For him the belief that external objects are mind-independent 'things' is a psychological problem rather than a philosophical question, as I shall discuss below.

Bachelard regarded realism, along with the other traditional epistemological doctrines, such as pragmatism and positivism, as rooted in nineteenth-century science.[18] The advancement of twentieth-century science for him called for a different set of philosophical questions. In his view, contemporary science has shown that it does not make sense to talk of scientific objects as if they were independent of the knower. In quantum mechanics, the measuring apparatus inevitably interferes with the system. In Niels Bohr's words, 'the procedure of measurements has an essential influence on the condition on which the very definition of the physical quantities in question rests'.[19] Bachelard argued that the only possible study of corpuscles is technical, that is to say it can only be done by using an experimental apparatus; in his own words 'of all corpuscles of modern physics, one can only do a phenomenotechnical study'.[20] For Bachelard, it is no longer a matter of observing

---

Meyerson in this article as the negative example, but the latter could not have been one of the philosophers who 'still' believed in the fixity of the mind, as he had died 16 years earlier.

15 Bachelard, *Le rationalisme appliqué*, pp. 9–10.

16 Between the late 1960s and the 1970s, a vigorous discussion took place among those who read Bachelard's philosophy as the result of an idealistic and even spiritualist tradition (M. Vadée, *Bachelard ou le nouvel idéalisme épistémologique* [Paris: Editions sociales, 1975]; Michel Serres, 'La réforme et les sept péchés', *L'Arc* 42 [1970]: 14–28) and those who read it as materialist (Dominique Lecourt, *Bachelard ou le jour et la nuit* [Paris: Grasset, 1974]). On the other hand, the sociologists Bourdieu, Chamboredon and Passeron included Bachelard and Canguilhem in the canon of sociology, along with Durkheim, Weber and Marx. See their volume designed for teaching students of the Ecole des hautes études en sciences sociales: Pierre Bourdieu, Jean-Claude Chamboredon and Jean-Claude Passeron, *The Craft of Sociology: Epistemological Preliminaries* (trans. Richard Nice; Berlin: de Gruyter, 1991 [1968]).

17 Gaston Bachelard, *L'activité rationaliste de la physique contemporaine* (Paris: Presses Universitaires de France, 1951), pp. 6–7.

18 Pariente has analysed the evolution of Bachelard's view of realism: Jean-Claude Pariente, 'Rationalisme et ontologie chez Gaston Bachelard', in Michel Bitbol and Jean Gayon (eds), *L'épistémologie française, 1830–1970* (Paris: Presses Universitaires de France, 2006), pp. 253–81.

19 Niels Bohr, 'Quantum Mechanics and Physical Reality', in John A. Wheeler and Wojciech H. Zurek (eds), *Quantum Theory and Measurement* (Princeton, New Jersey: Princeton University Press, 1983 [1935]), p. 144.

20 Bachelard, *L'activité rationaliste de la physique contemporaine*, p. 92. Bachelard employed the term '*corpuscules*' as a generic term for elementary particles. For a history of

phenomena, nor even a matter of creation of new concepts, as it ultimately was for Brunschvicg, but rather of doing a 'phenomenotechnical study' of them. 'Phenomenotechnique' is Bachelard's coinage, and it is aimed at indicating that scientific facts are not phenomena, that is to say they do not 'appear' to us, but they are rather technically defined, created and tested.[21] As he put it, for the scientific mind, 'nothing is given'; rather, 'everything is constructed'.[22] Bachelard directly criticized Edmund Husserl, the founder of phenomenology as a philosophical doctrine, for conceiving of knowledge as 'reception' of 'data' by the mind. He commented that Husserl's 'dualism' between mind and given was not 'close enough' and that the interaction between mind and object was not 'mutual' enough.[23] In modern physics, he argued, 'ontology [is] conditioned by technical experience'.[24]

The History of the Scientific Mind and the Timeless Epistemological Obstacles

In Bachelard's own words, he intended to show 'the mutual transformation of man and things'.[25] On the one hand, the mind for him transforms 'things', indeed it makes them 'rational', by 'rectifying' them, that is to say by removing their irregularities. On the other hand, he believed that in the process of knowledge, the subject also removes 'irregular attitudes from his intellectual behaviour'.[26] For Bachelard, it is scientific knowledge that changes the mind and that indeed gives it a history. In his account, the changes in ways of thinking are conquests that human beings make over their own innate tendencies. In *La formation de l'esprit scientifique*, Bachelard even

---

physicists' terms for elementary particles, see Helge Kragh, *Quantum Generations: A History of Physics in the Twentieth Century* (Princeton: Princeton University Press, 1999); table 13.1 (p. 197) records the old and new names of specific particles.

21 For discussions of Bachelard's concept of phenomenotechnique, see Teresa Castelão-Lawless, 'Phenomenotechnique in Historical Perspective: Its Origins and Implications for Philosophy of Science', *Philosophy of Science* 62 (1995): 44–59; Hans-Jörg Rheinberger, 'Gaston Bachelard and the Notion of "Phenomenotechnique" ', *Perspectives on Science* 13/3 (2005): 313–28; Cristina Chimisso, 'From Phenomenology to Phenomenotechnique: The Role of Early-Twentieth-Century Physics in Gaston Bachelard's Philosophy', *Studies in History and Philosophy of Science* (forthcoming).

22 Bachelard, *La formation de l'esprit scientifique*, p. 14.

23 Bachelard, *Le rationalisme appliqué*, p. 43. Bachelard's focus on the manipulation rather than observation of objects have suggested to some critics that his philosophy is close to Ian Hacking's, as he has presented it in *Representing and Intervening* (Ian Hacking, *Representing and Intervening* [Cambridge: Cambridge University Press, 1983]). See for instance M. Tijiattas, 'Bachelard and Scientific Realism', *Philosophical Forum* 22 (1991): 203–10; this view has been criticized: see Dan McArthur, 'Why Bachelard is not a Scientific Realist', *Philosophical Forum* 33/2 (2002): 159–72. I think it is very difficult to make Bachelard's epistemology fit into the categories of Anglo-American philosophy: his philosophy is not mainly aimed at answering questions such as whether unobservables exist independently of our mind.

24 Bachelard, *L'activité rationaliste de la physique contemporaine*, p. 82.

25 Ibid., p. 3.

26 Bachelard, 'Idéalisme discursif' [1934–35] in Gaston Bachelard, *Etudes* (Paris: Vrin, 1970), pp. 91–2.

proposed a quasi-Comtian three-stage model of historical development: the pre-scientific stage from classical antiquity to the eighteenth century; the scientific stage from the end of the eighteenth century to the beginning of the twentieth century; and finally the era of the new scientific mind, which started in 1905 with the publication of Einstein's theory of (special) relativity.[27]

La formation, however, is not mainly about the progress of the scientific mind, but about the obstacles that, in Bachelard's view, the mind has to overcome in order to become scientific. He provided a rather heterogeneous catalogue of them, including 'primary experience', general knowledge, over-extension of familiar images unitary and pragmatic knowledge, substantialism, realism, libido, the 'myth of digestion', and animism. These 'epistemological obstacles', as Bachelard called them, originate in the mind itself, and especially in the imagination and emotions, and guide our every-day experience, which becomes itself an obstacle to scientific knowledge when it is applied beyound the limits of our every-day life. 'Primary experience', that is all immediate and unreflective experience, which has not been rationalized and tested, is not just unscientific for Bachelard, it is indeed an obstacle to scientific knowledge.

In order to illustrate the epistemological obstacles, Bachelard employed many examples from the eighteenth century; one is that of the Abbé Poncelet who, in his 1796 book on thunder, dedicated a whole chapter to the fright that the thunder causes. For Bachelard such reflections belong to polite society, rather than science. The desire to possess riches is also for him rooted in human nature, indeed it is as natural as realism, which he described as 'the only innate philosophy that there is'. He argued that the two are in fact the same: realists are misers who want to possess the riches of reality; they suffer from the 'Harpagon complex'.[28] For Bachelard, libido is at the root of many pre-scientific theories: the alchemists often described their procedures as copulation and marriages, mercury as sterile, and advised that metals cannot be created from blood and human sperm. He argued that the philosophers of metals were also animist, as they attributed a soul to every being, and indeed regarded metals as living beings. Among his pieces of evidence, Bachelard cited the eighteenth-century Professor of Theoretical Chemistry and Director of the Prussian Royal Pharmacy Johann Heinrich Pott who reportedly detected many cases of 'mineral fertility'. For Pott, metals reproduce themselves, and as a consequence he was convinced that mines which had been exploited in ancient times, could be full of metals again.[29] Consistently, metals contracted diseases as well: Bachelard cited a seventeenth-century author who claimed that rust is a disease that attacks iron. Bachelard also quoted the German alchemist Johann Rudolph Glauber who argued that metals, when separated from the earth, are effectively separated from their source of nourishment, and that nature regulates the birth and death of metals just as she does for plants and animals.[30]

---

27 Bachelard, *La formation de l'esprit scientifique*, p. 7.

28 Ibid., ch. 7. Bachelard named the complex of the realist after Harpagon, the principal character in Molière's play *The Miser*.

29 Ibid., p. 158.

30 Ibid., p. 156.

Bachelard did not extract his last quotation from Glauber's work directly, but from Metzger's *Les doctrines chimiques*.[31] He repeatedly quoted Metzger's work when discussing seventeenth and eighteenth-century chemical doctrines.[32] This is hardly surprising, for she was the most prominent expert in the field. However, it is clear that their respective explanations of the ways of thinking of alchemists and philosophers of metals differed. As discussed in the previous chapter, Metzger aimed to demonstrate that their way of reasoning was neither absurd nor incorrect, in fact it was rational and coherent within a worldview that profoundly differed from that of modern science and of modern people in general. Bachelard explicitly wrote that the 'intuition of life' underlying animist theories was 'less intellectualist' than Metzger thought; in fact, it exhibited an 'affective character'.[33] Animist theories for him did not originate in rational thinking, but rather from emotions and desires. He thought that scientific mentality, on the one hand, and primitive, or 'pre-scientific', mentality, on the other, differ more profoundly than Metzger believed; indeed he assigned them to different mental faculties: rationality and imagination.

Bachelard argued that the alchemists' 'intuition of life' was not only less intellectualistic, but also more enduring than Metzger thought. Indeed, for him it was not specific to alchemic thought, but in fact it was detectable in many non-scientific texts, including recent ones.[34] He believed that there was an epistemological rupture between scientific knowledge and any other type of inquiry, and that non-scientific doctrines were expressions of the same way of thinking, rooted in human imagination and desires. Bachelard chose his examples of pre-scientific mentality from the most diverse sources, belonging to very different societies and times. He argued that in order to observe primitive approaches to phenomena, we do not need to limit ourselves to the observation of peoples who are considered primitive. For him, human beings face any new phenomenon in a 'primitive', that is, non-rational, manner: electricity was new in the eighteenth century, and as a consequence was immediately

---

31 Metzger, *Les doctrines chimiques en France du début du XVII*[e] *à la fin du XVIII*[e] *siècle*, p. 124, n. 1.

32 Bachelard quoted Metzger's *Les doctrines chimiques* elsewhere in *La formation* (pp. 51, 88, 145, 155, 156) and in other publications; see for instance: Gaston Bachelard, *Le pluralisme cohérent de la chimie moderne* (Paris: Vrin, 1973 [1932]), pp. 20, 44. He also quoted other works by her, including *Les concepts scientifiques* (Bachelard, *La formation de l'esprit scientifique*, p. 86) and Hélène Metzger, 'La philosophie de la matière chez Stahl et ses disciples', *Isis* 8 (1926): 427–64 (in Bachelard, *Le pluralisme cohérent de la chimie moderne*, p. 44). Bachelard's reference of Metzger's article has 1925 as year of publication, it is in fact in the 1926 volume.

33 Bachelard, *La formation de l'esprit scientifique*, p. 155. For some insight into Metzger's view of Bachelard's approach, see her reviews of his books: Hélène Metzger, 'Gaston Bachelard, *Essai sur la connaissance approchée*, Paris 1928; Gaston Bachelard, *La valeur inductive de la relativité*, Paris 1929; Gaston Bachelard, *Etude sur l'évolution d'un problème de physique*, Paris 1928', *Archeion* 12 (1930): 218–20; Hélène Metzger, 'Gaston Bachelard. *La Formation de l'esprit scientifique. Contribution à une psychanalyse de la connaissance objective*', *Archeion* 21 (1938): 162–5. In the latter, Metzger expressed her negative view of Bachelard's selective reading of past and indeed present science.

34 Ibid.

sexualized.[35] He also chose his sources of primitive thought from very diverse fields: in order to illustrate the sexual characters of many pre-scientific representations of fire, he did not only quote eighteenth-century natural scientists, but also Novalis, Gabriele D'Annunzio and Paul Valéry. He compared representations of fire in South-American and Australian myths with those made by European Romantic poets, and concluded that they were all rooted in images, desires and fears that do not belong to any society in particular, but to all human beings. Indeed, Bachelard read Novalis's poetry as 'an attempt to re-live *primitivity*'.[36]

His findings led him to discard 'sociological' explanations of 'primitive' practices and myths. He discussed in particular James Frazer's representation of fire festivals.[37] Frazer based his explanation of the Scottish fire festivals on the utility of bonfires that fertilize the fields. Bachelard responded that Frazer's interpretation was in fact an example of the 'unconscious rationalization' that a modern author, who knows about fertilizers, makes of behaviours which in fact originates in instincts and desires. The French sociological school, either in Durkheim's or Lévy-Bruhl's versions, could not satisfy Bachelard either. Mauss's comments during one of Lévy-Bruhl's sessions at the Société française de philosophie, cited in Chapter 3, give a clear idea of the reasons. Mauss argued that, despite their differences, he and Lévy-Bruhl were both sociologists because both believed that the mind has a history and that that history cannot be written without writing at the same time the history of societies. Bachelard believed that only the rational part of the mind has a history, while the emotional and imaginative part is timeless. Just as the historians of philosophy thought that philosophy enjoyed an 'eternal present', so, Bachelard thought, did the works of imagination. As a consequence, the history of societies would not have helped to explain pre-scientific thought. Rather, he believed that a new discipline, for which there was not yet a clear academic space, could explain the pre-scientific mind: psychoanalysis.[38] As he put it, he intended to show that it was necessary to 'correct' the 'sociological interpretation by a psychoanalytical interpretation'.[39]

35 Gaston Bachelard, *La psychanalyse du feu* (Paris: Gallimard, 1949 [1938]), pp. 53–4.

36 Ibid., p. 73; the English translation is from Gaston Bachelard, *The Psychoanalysis of Fire* (trans. Alan C.M. Ross; Boston: Beacon Press, 1964), p. 38.

37 Sir James George Frazer, *Myths of the Origin of Fire* (London: Macmillan & Co., 1930) and Sir James George Frazer, *The Golden Bough: A Study in Comparative Religion* (2 vols; London: Macmillan & Co., 1890). Along with Frazer's, Bachelard discarded the theories of the 'British school' as a whole.

38 Daniel Lagache, the pioneer of French academic psychoanalysts, obtained a psychology chair in Strasbourg in 1937 and in Paris in 1947: see Didier Anzieu, 'La psychanalyse au service de la psychologie', *Regards sur la psychanalyse en France. Nouvelle Revue de psychanalyse* 20 (1979): 59–75 (62). For a history of psychoanalysis in France, the classic works are Elisabeth Roudinesco, *Histoire de la psychanalyse en France. 1: 1885–1939* (Paris: Fayard, 1994 [1982]); Elisabeth Roudinesco, *La Bataille de cent ans: Histoire de la psychanalyse en France. 2: 1925–1985* (Paris: Seuil, 1986).

39 Bachelard, *La psychanalyse du feu*, p. 68; the English translation is from Bachelard, *The Psychoanalysis of Fire*, p. 35.

What for Lévy-Bruhl was primitive mentality for Bachelard was the unconscious as psychoanalysts described it.[40] Bachelard did not see primitive and modern mentalities as belonging to different people; indeed, he analysed modern mentalities[41] and found within them 'primitive' ways of thinking and acting. For him, both mentalities are present in modern people; 'even the scientist', he argued, 'when not practising his specialty, returns to the primitive scale of values'.[42] Indeed, since for Bachelard all our first encounters with phenomena are 'primitive', all knowledge begins in a primitive manner, or, as Bachelard put it, with reverie. In his words 'one can study only what one has first dreamed about'.[43]

Bachelard's acknowledgment that even science begins as reverie is very different from Metzger's view that 'primitive' ideas can be at the core of scientific theories, as for instance the idea of active analogy is for her at the core of the theory of universal attraction. For Bachelard it is not just a matter of enveloping a primitive idea in a self-critical and scientific framework, but rather of fighting and overcoming the instincts that are at the core of primitive ideas. For him psychoanalysis was not only an interpretative tool, but also a therapeutic device; indeed he declared his intention of 'curing the mind of its happy illusions'.[44] The subtitle of *La formation de l'esprit scientifique* is 'a psychoanalysis of objective knowledge': for Bachelard the 'primitive', emotional and imaginative elements of knowledge have to be recognized as such so that knowledge can become increasingly more objective.

Unlike the pre-scientific mind, the scientific mind for Bachelard has a history, in fact a rather eventful and dramatic history. Scientific knowledge in his view advances by endlessly criticizing what may seem already consolidated, in order to find new ways of ordering and understanding the world. New scientific knowledge for him emerges against instincts and first intuitions that we all carry within ourselves, and against old knowledge. In his words:

> The scientific mind must be formed *against* nature, against all that comes from nature's impetus and instruction, within us and outside us, against natural allurements and colourful, diverse facts. The scientific mind must be formed by being reformed.[45]

40 Bachelard's appropriation of psychoanalysis was rather personal, including as it did a Freudian terminology with references to Jung, and above all to the French psychoanalysts, such Marie Bonaparte and René Allendy. For a recent discussion of Bachelard's use of psychoanalysis, see Francesca Bonicalzi, *Leggere Bachelard: le ragioni del sapere* (Milano: Jaca Book, 2007), ch. 6, 'Psicoanalisi tra scienza e *rêverie*'.

41 So Bachelard described his aim in Bachelard, *La psychanalyse du feu*, p. 68.

42 Ibid., p. 16; the English translation is from Bachelard, *The Psychoanalysis of Fire*, p. 4.

43 Bachelard, *La psychanalyse du feu*, p. 48; the English translation is from Bachelard, *The Psychoanalysis of Fire*, p. 22.

44 Bachelard, *La psychanalyse du feu*, p. 16; the English translation is from Bachelard, *The Psychoanalysis of Fire*, p. 4.

45 Bachelard, *La formation de l'esprit scientifique*, p. 23; the English translation is from Gaston Bachelard, *The Formation of the Scientific Mind* (trans. Mary McAllester Jones; Manchester: Clinamen Press, 2002), p. 33.

In Bachelard's view, when human beings first turn their attention to an object, they cannot at first have a rational knowledge of it. As he put it, there are no 'first truths', but only 'first errors'. For him, it is only through a long process of rationalization of the object and of 'rectification' of the mind that scientific knowledge can emerge. The scientific mind for Bachelard must not only be 'educated', but indeed 'created'.[46] Unlike the pre-scientific mind, the scientific mind is for him a historical product, which emerges against its own instincts and indefinitely continues to 'reform' itself.

Bachelard argued that scientific knowledge needs not only a close engagement between the subject and the object, but also a dialectic exchange between minds. This is why he claimed that nobody could do science alone:[47] scientific work requires a bracketing of individuality; if individual concerns arise, they should be treated as problems and be addressed by psychoanalysis.[48] In his view, science is born of discussion and contradiction. The model of dialectical exchange that allows the advancement of knowledge for him is that of an ideal school, in particular of the exchanges that take place between teacher and pupil and between pupils themselves. He argued that pupils must be taught against their natural instincts, and older pupils should be 'monitors' of younger pupils. The need for 'surveillance' is central in Bachelard at all levels, including within the rational work of science, which for him needs a constant 'psychoanalysis'. He drew a parallel between the dialogue between teacher and pupil on the one hand, and the ego and super-ego within the individual subject on the other. Indeed he argued that Freud represented the super-ego as too authoritarian,[49] and proposed a super-ego based on objective values, rather than personal authority. The super-ego for him should not be the interiorization of the teacher's personal authority, unlike Freud's that results from the interiorization of parental authority. Indeed, Bachelard criticized parents and teachers for taking advantage of their authority, and for imposing their knowledge on young people.[50] For him criticism must go in both directions: from teacher to pupil and from pupil to teacher, and must be based on objective and rational values.[51]

---

46 'Idéalisme discursif' in Bachelard, *Etudes*.

47 See Bachelard, *Le rationalisme appliqué*, ch. 3, 'Rationalisme et co-rationalisme, l'union de travailleurs de la preuve'. Bachelard never changed his mind on this issue, and indeed he repeated it in his last writings, published posthumously: Gaston Bachelard, *Fragments d'une poétique du feu* (Paris: Presses Universitaires de France, 1988), p. 34.

48 Bachelard, *L'activité rationaliste de la physique contemporaine*, p. 4.

49 Bachelard's version of the splitting of the ego into ego and super-ego is certainly more positive and somewhat unproblematic when compared with Freud's. See for instance Freud's characterization of the super-ego in Sigmund Freud, *Civilization and its Discontents* (trans. David McLintock; London: Penguin, 2002 [1930]), pp. 60–61.

50 Bachelard, *Le rationalisme appliqué*, p. 75, see also the whole of ch. 4, 'La surveillance intellectuelle de soi'. In *Lautréamont*, Bachelard aimed to illustrate the effects that education that is not based on objective values can have on the minds of young people by examining the poet Isidore Ducasse (alias Lautréamont). See Gaston Bachelard, *Lautréamont* (Paris: Corti, 1986 [1939]), especially, ch. 3, 'La violence humaine et les complexes de la culture'.

51 I proposed my reading of Bachelard's rationalism as based on teacher–pupil dialectic in Chimisso, *Gaston Bachelard: Critic of Science and the Imagination*, ch. 3.

Bachelard did not want just to narrate the history of ways of thinking, or even to explain the reasons for their emergence, but also to advocate the necessity of change. He proposed a way in which the scientific mind can progress: by 'rectifying' itself, that is to say by progressively overcoming individual desires, and increasingly becoming more rational and objective. It is without doubt that Bachelard's proposal has a moral meaning and arguably moral overtones.[52] If the human mind does not change, for Bachelard is a matter of intellectual laziness. He commented that:

> We do not see this as proving the permanence and fixity of the human reason, as Meyerson thought, but rather as evidence of the somnolence of knowledge and the miserliness of cultivated minds that go over and over the same knowledge and culture and become, as all misers do, victims of the gold they so lovingly finger.[53]

A Split Mind and a Double Life

Despite his arguments of the moral value of scientific activity, Bachelard dedicated a great deal of his writings not only to the works of imagination, but also to the joys of reverie. Whereas in *La formation* and *La psychanalyse du feu*, Bachelard regarded instinctive attitudes and desires only as obstacles to scientific knowledge, in subsequent works he also analysed them independently of science. After his work on fire, he turned his attention to water, but already from the title of his book, it is clear that something has changed, as there is no mention of psychoanalysis. At the beginning of *L'eau et les rêves*, published four years after *La formation*, he declared that his objective was to 'become rationalist' in every respect, and that he had managed to do so about fire through a psychoanalysis of objective knowledge. However, he now confessed that the images of water still had a hold on him that rationality could not dispel or explain.[54] His study of images in poetry, reverie and in the life of imagination in general continued with works on the other two elements of ancient cosmology, earth and air.[55] These books are studies of the imagination, without the precise objective of helping rational thinking overcome it.

As he abandoned his original project to rationalize all images, he adopted a new phenomenological approach. In *La poétique de l'espace*, he explained his change of perspective: in his earlier work, following his habits as a 'philosopher of the sciences', he had interpreted images objectively, by avoiding his personal perspective.

---

52 In a classic article, Michel Serres has interpreted Bachelard's epistemological obstacles as deadly sins: covetousness (realism), libido, lust (sexualization of nature), sloth (non-science at large), pride (will to power and narcissism). He adds that Bachelard turned two deadly sins, envy and anger, into scientific virtues. Serres regards *La formation* as a whole as a work aimed at moral reform: Serres, 'La réforme et les sept péchés'.

53 Bachelard, *La formation de l'esprit scientifique*, p. 7; the English translation is from Bachelard, *The Formation of the Scientific Mind*, p. 19.

54 Gaston Bachelard, *L'eau et les rêves. Essai sur l'imagination de la matière* (Paris: Corti, 1942), p. 14.

55 Gaston Bachelard, *La terre et les rêveries de la volonté. Essai sur l'imagination de la matière* (Paris: Corti, 1992 [1947]); Gaston Bachelard, *La terre et les rêveries du repos. Essai sur les images de l'intimité* (Paris: Corti, 1992 [1948]); Gaston Bachelard, *L'air et les songes. Essai sur l'imagination du mouvement* (Paris: Corti, 1972 [1943]).

Now, he aimed to study the subjective character of images, and to consider them in 'their simplicity'.[56] In this book, as well as in *La poétique de la rêverie*, Bachelard studied images as they emerge in individual minds.[57] He employed his version of phenomenology for the images that populated our reveries. However, he believed that this type of phenomenological analysis was not applicable to scientific objects, which are never 'given', but always the result of both rationalization and technical transformation.[58]

The split of Bachelard's methodology reflects his double conception not only of objects, as images of our reveries and as scientific objects, but also of the mind. Indeed, he advocated a 'total separation between rational life and oneiric life', and proposed 'a double life' for human beings. On the one hand, for him there is rationality, which should overcome instincts and desires in order to produce scientific knowledge. The work of rationality is the work of the 'diurnal man', of social exchange and objectivity. On the other hand, there is imagination which feeds instincts and desires. The work of imagination is the work of the 'nocturnal man', of personal space and subjectivity.[59] In his view, rational knowledge and reverie at the beginning are confused, but soon proceed in opposite directions.[60] In order to carry out scientific work, an epistemological rupture with common knowledge is necessary.[61]

Bachelard's separation of rationality and imagination, and of science and reverie, is reflected in his works. In his books published in 1938, *La formation de l'esprit scientifique* and *La psychanalyse du feu*, he showed that images shaped by desire and instincts get mixed with objective knowledge and hinder its progress. In *La formation* in particular, he analysed how the study of nature had been distorted by the feelings and images that nature inspired in the more poetic part of our minds. In his books on current science, he analysed the works of rationality independently of the works of imagination.[62] Similarly, in his books on imagination, he became increasingly less concerned with objective knowledge.

---

56 Gaston Bachelard, *La poétique de l'espace* (Paris: Presses Universitaires de France, 1964 [1957]), pp. 2–4. Bachelard adopted the same phenomenological perspective in a subsequent book: Gaston Bachelard, *La poétique de la rêverie* (Paris: Presses Universitaires de France, 1960).

57 Bachelard, *La poétique de l'espace*, p. 3.

58 See Gaston Bachelard, *Le matérialisme rationnel* (Paris: Presses Universitaires de France, 1972 [1953]), p. 24.

59 See ibid., p. 19.

60 Bachelard, *La psychanalyse du feu*, p. 12.

61 Bachelard, *Le rationalisme appliqué*, ch. 4, 'Connaissance commune et connaissance scientifique'; see also Bachelard, *Le nouvel esprit scientifique*, pp. 140–41; Bachelard, *L'activité rationaliste de la physique contemporaine*, p. 93.

62 He published books on science before and after he developed his interest in imagination: Bachelard, *Essai sur la connaissance approchée* [1927]; Bachelard, *Etude sur l'évolution d'une problème de physique* [1927]; Gaston Bachelard, *La valeur inductive de la relativité* (Paris: Vrin, 1929); Bachelard, *Le pluralisme cohérent de la chimie moderne* [1932]; Bachelard, *La philosophie du non* [1940]; Bachelard, *L'activité rationaliste de la physique contemporaine* [1951]; Bachelard, *Le matérialisme rationnel* [1953].

In terms of historical examples, his books on images and on science are very different. Whereas in the former he could choose examples from any time and situation, for he believed that images are ultimately rooted in a part of the mind that does not change, in his books on science he only focused on relatively recent developments of scientific disciplines. Before the nineteenth century for him science did not exist; as a consequence a history of sixteenth-century 'science' would have been a misnomer. Moreover, in his view, unlike the pre-scientific mind, the scientific mind has not only a history, but a discontinuous history. He believed that there is an epistemological rupture not only between scientific and common knowledge, but also between different stages of scientific knowledge. Indeed, for him scientific knowledge, in order to advance, needs to 'say no' to old knowledge.[63] As a consequence, in his view, twentieth-century science was the expression of 'the new scientific mind', which differed from the scientific mind of the previous century.

Bachelard separated 'primitive', or pre-scientific, thought from modern thought at least as sharply as Lévy-Bruhl did, as he assigned them to different mental faculties. For Bachelard reason and imagination are opposed to each other; the former gives rise to science, which is a social activity; while the latter to reverie and poetry, which for him are individual activities. However, unlike Lévy-Bruhl, he believed that these two ways of approaching reality are present in modern minds, and indeed that human beings need both, as they need rationality as well as affectivity, knowledge as well as dreams. Brunschvicg's history of the mind as progressive rationalization becomes only half of the story for Bachelard: he accepted it as far as science was concerned, but assigned great importance to the joys of being primitive in one's private dreams. For him, the primitive mind lives within us, although it should not influence scientific knowledge.

Unlike Metzger and Koyré, Bachelard did not only aim to describe doctrines and ways of thinking, but rather to evaluate whether they are scientific. He was concerned with sorting science from non-science, and the history of science from the history of non-scientific activities. Indeed, for him, strictly speaking, Metzger's and Koyré's works were not about the history of science, for they studied periods in which in his view science had not yet emerged. All of them aimed to understand ways of thinking in their own terms, but Bachelard went further: he judged them, and declared them scientific or non-scientific. Naturally, in order to judge, he needed a norm, and this norm was current science. Michel Foucault clearly described Bachelard (and Canguilhem's) 'epistemological history':

> A type of historical analysis … [that] takes as its norm the fully constituted science; the history that it recounts is necessarily concerned with opposition of truth and error, the rational and the irrational, the obstacle and fecundity, purity and impurity, the scientific and non-scientific. It is an epistemological history of science.[64]

---

63 Bachelard described his philosophy as 'the philosophy of no', as he explained in his *La philosophie du non*. The theme of epistemological discontinuity was already present in his early works: Bachelard, *Essai sur la connaissance approchée*, p. 270.

64 Michel Foucault, *The Archaeology of Knowledge* (London: Tavistock, 1972 [1969]), p. 190.

Bachelard's defence of scientific rationality and of modernity was clearly consistent with Brunschvicg's philosophy. However, there are important differences between them. Partly for obvious chronological reasons, in Bachelard's work the reader does not find the defence and promotion of secularism, which was important for his Third Republic teacher. The early parts of Brunschvicg's career developed in the charged environment of the Dreyfus affair and its aftermath, which gave great political meaning to being on the side of secularism and republicanism, and of progress as opposed to tradition. Bachelard's philosophy had more moral than political aims, focused as it was on the necessity for the scientific mind to overcome individual interests, dreams and images. Bachelard did not have the political commitment that Brunschvicg, not to mention Lévy-Bruhl, had. Bachelard further worked on the themes and issues, such as mentality, rationality and progress, which had been at the core of the thought of inter-war philosophers such as Lévy-Bruhl, Brunschvicg and Rey. However, their philosophies appeared to be more in contact with the political and ethical issues of their time than Bachelard's was. The latter's philosophy maintained important links with the philosophies of the previous generation of Sorbonne philosophers, but it was less engaged than theirs with the social and political events of its time, which stretched from the late 1920s to the early 1960s.

Georges Canguilhem between Concepts and the Living Being

A Scholar between Two Worlds

Georges Canguilhem's life spanned from 1904 to 1995, covering almost the whole of the twentieth century. His philosophical and ethical education took place in the Third Republic, and his epistemological work bears the signs of continuity with questions, aims and methodologies of the programmes of the history of the mind that I have discussed in this book. A pupil of Alain at the *lycée* Henri-IV and a contributor to his teacher's publication *Libres propos*,[65] he continued his education at the Ecole normale supérieure, along with Jean-Paul Sartre, Raymond Aron and Paul Nizan; Merleau-

---

65 For a discussion of Canguilhem's articles published in *Libres propos* and of his early (pre-1943) work in general, see Jean-François Braunstein, 'Canguilhem avant Canguilhem', *Revue d'histoire des sciences* 53/1 (2000): 9–26. In the 1930s, Canguilhem started to dissociate himself from Alain and his followers. He particularly objected to their enduring pacifism, which lead Alain to support the Munich agreement, and later some of his followers even to collaborate with the Nazis. As Braunstein has emphasized, Canguilhem clearly understood that the situation was no longer comparable to the First World War: Braunstein, 'Canguilhem avant Canguilhem', p. 15. Canguilhem's 'prudent' teaching style, careful about historical accuracy, has been contrasted with the dazzling improvisations of his teacher Alain: see Bertrand Saint-Sernin, 'Georges Canguilhem à la Sorbonne', *Revue de métaphysique et de morale* 90 (1985): 84–92 (84). By contrast, Daniel Lagache, in his review of *Le normale et le pathologique*, published in 1946, thought that in the style of the book, the reader would find 'Alain's former pupil': Daniel Lagache, 'Le normal et le pathologique d'après M. Georges Canguilhem', *Revue de métaphysique et de morale* 51 (1946): 355–70 (369).

Ponty entered the Ecole one year later.[66] His prestigious Parisian education was far more fitting for a Sorbonne professor, which he became in 1955, than that of his predecessor Gaston Bachelard. However, as in the case of Bachelard, his provincial, indeed rural, origins have often been commented on, even by himself. Despite being the son of a tailor, he often said that he was of peasant stock.[67] His geographical and social roots made him stand out in the academic milieu, hence comments on his accent and body language,[68] on his social difficulties at boarding school, and on the contrast between himself and his fellow students at the Ecole normale, exemplified, in Bourdieu's account, by their different sport activities: Canguilhem played rugby whereas Sartre and Aron played tennis at a very high level.[69] His career also appeared to follow a familiar path: he first taught in provincial *lycées*, and then in a provincial university (Strasbourg, which had relocated to Clermont-Ferrand, as the city of Strasbourg had been annexed by the Third Reich). After the war, he

66 For his *diplôme d'études supérieures*, Canguilhem wrote a thesis under the supervision of Célestin Bouglé, on 'La théorie de l'ordre et du progrès chez Auguste Comte' in 1926. Jean Cavaillès gained his *diplôme* in the same year: see *Annales de l'Université de Paris*, (1927), p. 299. See also François Dagognet, *Georges Canguilhem: philosophe de la vie* (Le Plessis-Robinson: Institut Synthélabo, 1997), p. 9. Jean Cavaillès, mathematician and philosopher, founded the journal *Dialectica* together with Bachelard. During the war, he played a prominent role in the Resistance; he was captured and killed by the Nazis in 1944. Canguilhem's speeches in his honour are collected in Georges Canguilhem, *La vie et mort de Jean Cavaillès* (Paris: Allia, 1996 [1976]).

67 Paul Rabinow, 'Introduction: A Vital Rationalist', in François Delaporte (ed.), *A Vital Rationalist: Selected Writings from Georges Canguilhem* (New York: Zone Books, 1994), p. 11. Indeed, interviewed in 1995, the year of his death, he still mentioned his rural origin, see François Bing and Jean-François Braunstein, 'Entretien avec Georges Canguilhem', in François Bing, Jean-François Braunstein, and Elisabeth Roudinesco (eds), *Actualité de Georges Canguilhem. Le normale et le pathologique* (Paris: Synthélabo, 1998), pp. 121–35 (135). Canguilhem described his younger self as a student of the Ecole normale supérieure from Languedoc, who studied towards his *agrégation*, and spent the rest of the time working in the fields; see Braunstein, 'Canguilhem avant Canguilhem', p. 11. See also Claude Debru, *Georges Canguilhem, science et non-science* (Paris: Editions rue d'Ulm/Presses de l'Ecole normale supérieure, 2004), p. 15; and Dominique Lecourt, 'Georges Canguilhem, le philosophe', in Jean-François Braunstein (ed.), *Canguilhem: Histoire des sciences et politique du vivant* (Paris: Presses Universitaires de France, 2007), pp. 27–43 (27).

68 See for instance Pierre Bourdieu, 'Georges Canguilhem: An Obituary Notice', *Economy and Society* 27/2–3 (1998): 190–91; David Macey, 'The Honour of Georges Canguilhem', *Economy and Society*, 27/2–3 (1998): 171–81 (175).

69 Bourdieu, 'Georges Canguilhem: An Obituary Notice', p. 191. For the meaning of Bourdieu's contrast between rugby and tennis, the best explanation is given by Bourdieu's own analysis: he argued that sports like football and rugby in France 'combine all features which repel the dominant class: not only the social composition of their public … but also the values and virtues demanded, strength, endurance, violence, "sacrifice", docility and submission to collective discipline … [By contrast] all the feature which appeal to the dominant taste are combined in sports such as golf, tennis, sailing, riding': Pierre Bourdieu, *Distinction: A Social Critique of the Judgment of Taste* (London: Routledge, 1989 [1979]), pp. 215, 16.

served as philosophy *inspecteur général*[70] before succeeding Gaston Bachelard at the Sorbonne in 1955. This list of his professional activities hides the dramatic events that marked his early career, including his resignation from his *lycée* teaching post in 1940, because, as he put it, he did not want to teach the doctrine of Marshal Pétain, and his participation in the Resistance.[71]

The world in which he developed his career and published his work, however, was very different from that in which Lévy-Bruhl, Brunschvicg, Metzger and Rey published theirs.[72] In addition to the macroscopic political and social changes, profound changes also took place in the world of philosophy, both socially and theoretically. A new type of philosopher as independent and public intellectual *à la* Sartre was replacing the Third Republic 'little teacher',[73] for whom the classroom and academic publications were the main spaces and means of his social engagement. Canguilhem, however, appeared to resemble more closely a Third Republic mandarin than a 1960s *engagé* philosopher. Indeed, he defended the role of the philosopher as teacher rather than as a writer involved in current affairs.[74] Bourdieu has explained the fact that Canguilhem was not a public intellectual in terms that leave no doubt that he sided with him: while regretting that Canguilhem left the spotlight to 'show-offs and impostors', he judged his style as incompatible with 'the mystical-literary enthusiasms for the existential exaltation of Hölderlino-Heideggerian thought which enchants the poet-thinker'.[75]

The content of Canguilhem's research was indeed different from the type of philosophy that after the war almost came to be synonym with French philosophy, namely existentialism. Michel Foucault drew a clear opposition between Bachelard, Koyré and Canguilhem's style of philosophy, which engages with knowledge, rationality and history, on the one hand, and Sartre and Merleau-Ponty's philosophies, which are concerned with experience and the subject, on the other.[76] Canguilhem continued the philosophical tradition of reflection on the sciences, of which Brunschvicg and Bachelard were the most eminent exponents.

---

70 He replaced George Davy (whom we have met in the previous chapters as a sociologist) as philosophy *inspécteur général* in 1947, after refusing the post immediately after the war, see Jacques Lautman, 'Un stoïcien chaleureux', *Revue d'histoire des sciences* 53/1 (2000): 27–45 (34).

71 Bing and Braunstein, 'Entretien avec Georges Canguilhem', pp. 122–4; see also: Rabinow, 'Introduction: A Vital Rationalist' in Delaporte (ed.), *A Vital Rationalist: Selected Writings from Georges Canguilhem.*

72 Although he co-authored a handbook before the fall of the Third Republic (Georges Canguilhem and Camille Planet, *Traité de Logique et de Morale* [Marseille: Robert & Fils, 1939]), his most important works were published after it.

73 Cf. Fabiani's depiction of Third Republic philosophers as academics, who may even be seen as 'little teachers' rather than great minds: Fabiani, *Les philosophes de la République*, especially Introduction and ch. 1.

74 See Guillaime Le Blanc, *Canguilhem et le normes* (Paris: Presses Universitaires de France, 1998), pp. 22–7.

75 Bourdieu, 'Georges Canguilhem: An Obituary Notice', p. 191.

76 Michel Foucault, 'La vie: l'expérience et la science', *Revue de métaphysique et de morale* 90 (1985): 3–14 (4, 14).

Like Bachelard and Brunschvicg, Canguilhem had both philosophical and scientific training. Just as Bachelard worked towards his doctorate in philosophy while teaching sciences in a secondary school, so Canguilhem wrote his medical dissertation while teaching philosophy in a *lycée*. However, their sciences were not the same. Already differences could be detected between Brunschvicg's major focus on mathematics, which made his idealism look all the more natural, and Bachelard's interest in chemistry and physics, which stimulated him to attribute central importance to the technical part of science.[77] Canguilhem's choice to concentrate on the life sciences and medicine lent a novel perspective to history and indeed epistemology. What philosophical consequences follow from his choice? Can the concept of scientific object be transferred from physics to medicine? Does his view of discontinuities in the history of science as partial and complex derive from his study of the life sciences as François Dagognet has argued? As we shall see, Canguilhem accepted Bachelard's normative view of the history of science, but can the production of norms be the same in such different sciences?

Historical Epistemology and the History of Concepts

Canguilhem presented his view of the relationship between philosophy and history as substantially inherited from Bachelard.[78] His critical target was the positivist models of the history of science, which proposed a progressive narrative without, in his view, properly evaluating past theories and concepts. Like Bachelard, he believed that only by adopting a normative approach can the historian distinguish which doctrines are connected with present-day science and which are not. He expressed his conception of a double history of science by recalling Bachelard's distinction between 'sanctioned' and 'lapsed' history of science. The former is the history of the doctrines that current science has absorbed (or sanctioned), even if in a transformed manner, and the latter is the history of those doctrines which have fallen out of scientific research.[79] This model might sound like a revised version of the progressive and anachronistic model of history, for it reconstructs a narrative

77 For the importance of chemistry in particular in Bachelard's work, see Bensaude-Vincent, 'Chemistry in the French Tradition of Philosophy of Science: Duhem, Meyerson, Metzger and Bachelard'.

78 See Georges Canguilhem, 'Le rôle de l'épistémologie dans l'historiographie scientifique contemporaine', in Georges Canguilhem, *Idéologie et rationalité dans l'histoire des sciences de la vie* (Paris: Vrin, 1993 [1977]). I have compared Bachelard's and Canguilhem's views of normative history in Cristina Chimisso, 'The Tribunal of Philosophy and its Norms: History and Philosophy in Georges Canguilhem's Historical Epistemology', *Studies in History and Philosophy of Biological and Biomedical Sciences* 34/2 (2003): 297–327.

79 Georges Canguilhem, 'L'objet de l'histoire des sciences' [1966], in Canguilhem, *Etudes d'histoire et de philosophie des sciences concernant les vivants et la vie*, p. 13. See also Georges Canguilhem, 'L'histoire des sciences dans l'œuvre épistémologique de Gaston Bachelard' [1963], in Canguilhem, *Etudes d'histoire et de philosophie des sciences concernant les vivants et la vie*, p. 177. For Bachelard's concepts of lapsed and sanctioned history, see Bachelard, *L'activité rationaliste de la physique contemporaine*, pp. 21–31. I have discussed

from the point of view of the present. Canguilhem was aware of this possible interpretation, which he explicitly rejected.[80] Like Bachelard, he did not use current science in order to reconstruct uncritical continuities, nor did he read past doctrines as anticipating present-day science. He discarded this teleological model, and added his voice to that of previous scholars, including Metzger and Koyré, in condemning the absurdity of the concept of 'precursor', which was at the very core of that model.[81] Again similarly to Bachelard, Canguilhem employed epistemology in order to judge whether past doctrines were in fact related to current ones, or whether the continuities that had been traditionally reconstructed were untenable.[82]

In Canguilhem's view, without a normative point of view, it is not possible for the historian of science to construct any history. He thought that it is not only a matter of selecting theories and events that should have a place in a narrative, but also of having a guide on how to connect them correctly. When historians of science set out to write the history of a particular discipline, how do they decide which doctrines belong to it? From a purely historical point of view, the science under scrutiny is limited to the field determined by past scholars. However, if historians aim to construct a narrative or a genealogy of doctrines, the purely historical perspective for Canguilhem can lead to serious mistakes. He provided the example of botany. A historian might write a history of botany by examining the research that, at any given time, went under this name. However, Canguilhem explained that eighteenth-century botanists based the physiology of plants on that of animals, while modern botanists rely on chemistry and physics. It would not be prudent, he concluded, to establish a continuity between eighteenth-century projects and modern botany, and to hide the 'radical discontinuity' and the 'radical novelty' of biochemistry and biophysics.[83]

Bachelard attacked the unreflective continuities of positivistic history of science by showing the epistemological ruptures that separated the supposed past of physics and chemistry from the modern disciplines. On his part, Canguilhem paid more attention to the possibility of constructing continuities. As mentioned, François Dagognet has interpreted Canguilhem's view of discontinuities as partial

these concepts in Chimisso, *Gaston Bachelard: Critic of Science and the Imagination*, pp. 92–106.

80 Georges Canguilhem, 'L'objet de l'histoire des sciences' [1966], p. 14.

81 Georges Canguilhem, 'L'objet de l'histoire des sciences' [1966], pp. 20–23. Canguilhem here quoted Koyré in his support. On this point, see Hans-Jörg Rheinberger, 'Reassessing the Historical Epistemology of Georges Canguilhem', in Gutting (ed.), *Continental Philosophy of Science*, pp. 191–2. See also Hélène Metzger, 'Le rôle des précurseurs dans l'évolution de la science', *Thalès* 4 (1939): 199–209, reprinted in Metzger, *La méthode philosophique en histoire des sciences*. On Canguilhem's criticism of the notion of precursor, see also Gilles Renard, *L'épistémologie chez Georges Canguilhem* (Paris: Editions Nathan, 1996), pp. 52–9.

82 See Canguilhem's presentation of Bachelard's view according to which epistemologists should retrace the development of scientific thought by 'judging' the documents gathered by historians: Georges Canguilhem, 'L'histoire des sciences dans l'œuvre épistémologique de Gaston Bachelard', in Canguilhem, *Etudes d'histoire et de philosophie des sciences concernant les vivants et la vie*, p. 177.

83 Canguilhem, *Idéologie et rationalité dans l'histoire des sciences de la vie*, p. 15.

and complex as a consequence of his focus on the life sciences. These, he argued, are at an earlier stage of development than physics and chemistry, and therefore their development is slower and more continuous. Indeed, Canguilhem himself, citing Bachelard, argued that a 'continuist' history of science is a history of 'young' science.[84] Although Dagognet certainly makes a good point, it is nonetheless striking that Canguilhem went back centuries to find concepts that for him could be connected to current ones.

Canguilhem offered a detailed example of his method in his history of the concept of reflex movement, in which he used epistemology in order to construct new narratives, and to reject those currently provided by historians. In *La formation du concept de réflexe aux XVIIe et XVIIIe siècles*, Canguilhem set out to demonstrate that the connection between Descartes's and the modern concepts of reflex movement was untenable. In Descartes's physiology, Canguilhem explained, the flux of 'spirits' in involuntary body movements is always from the brain towards the periphery, and never in the opposite direction. As a consequence, Descartes's reflex movement is qualitatively different from that of modern physiology, in which the movement is triggered by a stimulus of the peripheral nervous system, which is transmitted to the spinal chord, as in the knee-jerk reflex.

Having established this epistemological rupture within what was normally seen as a continuous narrative, Canguilhem, however, set up a new continuity. Indeed, he argued that a first concept of reflex, as modern science understands it, was proposed by Thomas Willis (1621–75), Professor of Natural History at Oxford, and of Medicine in London. Willis's concept was part of his rather imaginative theory in which he interpreted life as light: he employed optical laws of reflection in the interpretation of biological phenomena, so allowing for the modern concept of reflex movement.

With his example of the history of the concept of reflex, Canguilhem aimed to prove that a simple historical approach to the history of science could not give us a truthful narrative. The 'logic of history', he argued, may suggest that the modern concept of reflex had its origins in a mechanistic theory, namely Descartes's. However, he continued, a proper analysis of Descartes's theory shows that there is no connection between his and the modern concept of reflex. Canguilhem concluded that as far as the history of science is concerned, 'the rights of logic should not be replaced by the rights of the logic of history'.[85] He also used his history of the emergence of the concept of reflex in order to disprove what he saw as a prejudice, namely that a concept employed in a current scientific theory could not have emerged within a theory which is now deemed non-scientific.[86]

---

84 Georges Canguilhem, 'L'objet de l'histoire des sciences' [1966], in Canguilhem, *Etudes d'histoire et de philosophie des sciences concernant les vivants et la vie*, p. 14. Canguilhem explained Koyré's rather continuist approach also with the fact that he studied early science. Elsewhere, however, he criticized Koyré for exaggerating the philosophical discontinuity caused by the scientific revolution, as I shall discuss below.

85 Georges Canguilhem, *La formation du concept de réflexe aux XVIIe et XVIIIe siècles* (Paris: Presses Universitaires de France, 1955), p. 5.

86 Ibid., p. 3.

Canguilhem's defence of a history of concepts partially independent of theories and indeed metaphysical assumptions introduces a new historiographical perspective. The historians of philosophy and of science whom I have discussed in this book regarded concepts (and indeed theories) as dependent on general worldviews, mentalities and metaphysical assumptions. Their subject matters are different, but a shared approach is apparent: Lévy-Bruhl interpreted 'primitive' beliefs as consequences of a particular logic and worldview, just as Brunschvicg judged Kantian categories to be derived from the Newtonian worldview.[87] Analogously, Metzger explained the alchemists' research programme of turning one metal into another by their worldview dominated by 'active analogy'. Koyré interpreted Galileo's revolution as brought about by the latter's rejection of an Aristotelian approach to nature in favour of an Archimedean and Platonic one. Canguilhem too thought that metaphysical assumptions and worldviews shaped concepts. However, he not only believed that changes of worldviews took place in a slow and fragmentary manner, but also and crucially that concepts could survive within different worldviews, metaphysical assumptions and indeed theories. Willis's concept of reflex may have depended on his view of life as light, but for Canguilhem it stood in a relation of continuity with the current concept, despite the sharp contextual theoretical and metaphysical discontinuities.

Although Koyré did represent the scientific revolution as a long process, Canguilhem nonetheless believed that his representation of Galileo's 'conversion' to Platonism was too clear-cut and general. Indeed, he suspected that Koyré's reading was inspired by modern views, rather than by a historically accurate exegesis. Canguilhem saw less discontinuity between Galileo's and Aristotelian views of nature, and argued that in fact Galileo kept important aspects of the Aristotelian tradition. Beyond this specific interpretative disagreement, Canguilhem's work does exhibit a rather fragmented and complex view of scientific change. For him the scientificity of a concept, or its potential scientific value, does not appear to depend on his general assessment of theories, let alone worldviews and mentalities, in which it emerges.[88]

In this respect, Canguilhem's approach also differs from Bachelard's, notwithstanding their philosophical similarities. Bachelard would have never accepted as scientific any part of a theory as imaginative as that of Willis's, which indeed would have provided plenty of material for his psychoanalysis of knowledge. Indeed, their respective concepts of 'scientific' do not neatly overlap. Canguilhem explained that the judgements of his epistemological tribunal were neither purges nor executions. In other words, his aim was not so much to show that certain 'methods' and 'attitudes' were not scientific, but rather to show that they merit to

87 Canguilhem recalled Brunschvicg's thesis (without mentioning Brunschvicg) when he argued that Kant was following the culture of his time when he believed that he could abstract his table of categories, which in his judgement were absolute, from the science of his time; if, Canguilhem continued, one conceives of the history of science in terms of progress of enlightenment, it is difficult to conceive of the possibility of a history of the categories of scientific thought: Canguilhem, *Idéologie et rationalité dans l'histoire des sciences de la vie*, p. 20.

88 Canguilhem, *Idéologie et rationalité dans l'histoire des sciences de la vie*, p. 25.

be called scientific even if they have been superseded by new ones, insofar as they themselves in their time had superseded previous ones.[89] Bachelard would have accepted Canguilhem's view within what for him was science, which however went back only about two centuries and would not have included a seventeenth-century scholar. Their respective analyses were also at different levels of generality. Whereas Canguilhem focused on concepts, Bachelard explained discontinuities not only at the level of theories, but indeed at the level of ways of reasoning.

Bachelard's and Canguilhem's applications of norms had different scopes and levels of generality, but both their approaches were certainly normative. In *La formation du concept de réflexe*, Canguilhem judged Willis's and Descartes' respective concepts of reflex by using the modern concept of reflex as his norm. His epistemological history was reflexively normative, for he consciously employed current science as a norm to judge past knowledge. He aimed to demonstrate that epistemology is needed in order to create accurate narratives. These narratives are of course accurate from the point of view of current science, but reflexively so. A little like the theory of relativity, in which the observer cannot be eliminated, in Canguilhem's epistemological history, the historian acquainted with modern science is openly present. Its explicit perspective distinguishes it from positivistic histories of science, which presented a narrative without an observer, and therefore as absolutely objective. Like Bachelard, Canguilhem was well aware that observers, and therefore narratives, could change and would change; science for him is a historical product, and at different times there would not only be different sciences, but different perspectives of the past.

In *La formation du concept de réflexe*, Canguilhem used epistemology in order to construct new narratives, whereas Bachelard employed the history of science in order to elaborate a new epistemology. This difference has prompted some illustrious critics to label Canguilhem's project 'epistemological history' and Bachelard's 'historical epistemology'.[90] Michel Fichant proposed a more complex categorization, agreeing

---

89 G. Canguilhem, 'L'objet de l'histoire des sciences' [1966], in Canguilhem, *Etudes d'histoire et de philosophie des sciences concernant les vivants et la vie*, p. 14.

90 See Lecourt 'For a Critique of Epistemology' [1972], published in Lecourt, *Marxism and Epistemology. Bachelard, Canguilhem and Foucault*, p. 166; Jean Gayon, 'The Concept of Individuality in Canguilhem's Philosophy of Biology', *Journal of the History of Biology* 31 (1998): 205–325 (307, n. 8); Rheinberger has investigated the concepts of historical epistemology and epistemological history, focusing on Canguilhem's post-1960 works: see Rheinberger, 'Reassessing the Historical Epistemology of Georges Canguilhem'. As mentioned above, Foucault discussed Bachelard and Canguilhem's '*epistemological* history of the sciences' (original emphasis) in *The Archaeology of Knowledge*: Foucault, *The Archaeology of Knowledge*, p. 190. Althusser too implicitly judged Canguilhem's work more historical than Bachelard, when he declared that new directions of research had been opened by Cavaillès, Bachelard and Vuillemin in the philosophy of science, and by Canguilhem and Foucault in the history of science: Louis Althusser, 'Pierre Macherey, "La philosophie de la science de Georges Canguilhem": Présentation', *La Pensée* 133 (1964): 50–54 (51). Canguilhem called Thomas Kuhn's philosophy 'historical epistemology': 'Le statut épistémologique de la médicine', in Canguilhem, *Etudes d'histoire et de philosophie des sciences concernant les vivants et la vie*, p. 426.

on crediting 'historical epistemology' to Bachelard (and Cavaillès), but ascribing epistemological history to Koyré, and epistemology of the biological sciences to Canguilhem.[91] These categorizations show that there is a complex continuum which contains many of the projects discussed in this book, some more focused on history, but still with epistemological concerns and often aims, others more focused on the elaboration of new epistemologies, but at the same time based on historical analyses. I have already discussed some of the tension that characterized this group of projects when it came to emphasize history or epistemology. There is no doubt that Canguilhem too, despite his focus on history, could be seen as a philosopher rather than a historian. History and epistemology are intimately connected in Canguilhem's work, and each is necessary to the other. Whereas the use of epistemology in order to create a new history of science is apparent in the *La formation du concept de réflexe aux XVIIe et XVIIIe siècles*, the use of history in order to create a new epistemology is at the forefront in his other book, *Le normal et la pathologique*.

The History of Norms and the Individual

In *La formation du concept de réflexe*, Canguilhem showed that epistemological analysis created different narratives from those that current historiography had established. In *Le normale et le pathologique*, he both wrote a history of concepts, and proposed to revise epistemological concepts following the result of this history. Epistemological history and historical epistemology appear to converge in this book.[92] But there is more: the concepts that he examined historically included those of 'normal', 'abnormal', and indeed that of 'norm', which of course is at the core of the concepts of normal and pathological. As a consequence, Canguilhem did not only write a normative history, but a history of normativity. However, in the Introduction, Canguilhem presented his monograph as a philosophical rather than a historical work. Indeed, he explicitly wrote that he had not aimed to write a history of medicine, and that he had offered a historical presentation of his material only for the sake of clarity. His aim was rather to revise a thesis, generally accepted in the nineteenth century, that the pathological is a quantitative variation of the normal state. Appealing to the authority of Brunschvicg, who said that philosophy was the science of solved problems, he declared that he wanted to re-open a discussion rather than close it.[93]

Canguilhem described the first part of his book as research into historical sources, and the second part as the critical presentation of his own doctrine. Indeed, in the first part he particularly focused on August Comte's, Claude Bernard's and

---

91 Michel Fichant, 'L'épistémologie en France', in François Chatelet (ed.), *Histoire de la philosophie: le 20e siècle* (Paris: Hachette, 1973), pp. 135–78 (168–71).

92 *Le normale et le pathologique*, as published today, has two parts: the first is the 1943 *Essai sur quelques problèmes concernant le normal et le pathologique*, which originally was Canguilhem's doctoral thesis in medicine; the second part, *Nouvelles réflexions concernant le normal et le pathologique (1963–66)*, was added by Canguilhem to the 1966 edition.

93 Georges Canguilhem, *Le normal et le pathologique* (Paris: Presses Universitaires de France, 1999 [1966]), pp. 7–9.

René Leriche's ideas. He gave particular attention to Claude Bernard (1813–78), who tirelessly insisted on the continuity between physiological and pathological phenomena.[94] A clear example of this interpretation is Bernard's view of diabetes. The presence of sugar in the bloodstream is a normal physiological phenomenon; in the diabetic the quantity of sugar is higher than 'normal': the pathology is therefore a quantifiable variation of a normal state.[95] Canguilhem judged Bernard's solution of the problem of the relationship between the normal and the pathological similar to that offered by Comte. For Canguilhem, both philosophies bore the mark of nineteenth-century optimism, which did not attach any reality to evil.[96] As a consequence, for nineteenth-century physicians, illness has no independent ontological reality, but it is simply a quantitative variation of the normal state.

Canguilhem established both continuities and discontinuities between the positivistic conceptions of pathology as a quantitative variation of the normal state, and previous medical and philosophical ideas.[97] On the one hand, he saw a fundamental discontinuity between eighteenth-century and nineteenth-century medical theories, as the latter broke with the previously held dualist, indeed 'Manichaeistic', view according to which 'Health and Disease fought over man the way Good and Evil fought over the World'.[98] On the other hand, Canguilhem regarded nineteenth-century medical doctrines as the result of ideas that had been emerging since the Renaissance. Noteworthy among these was the Baconian view that human beings could and indeed must manipulate nature so that a desirable state of affairs could be restored; in the case of medicine, this was health.[99] Just as, from a Baconian perspective, the study of nature would empower human beings to obtain what they desire, so in nineteenth-century medicine, the study of physiology (as the 'normal' state to be restored) would enable medics to grapple with pathology, which in turn provided the basis for therapeutics.

Canguilhem's discussions of past doctrines is firmly aimed at the solution of philosophical questions. He was chiefly interested in establishing whether the pathological is merely a quantitive modification of the normal state, and whether sciences of the normal and the pathological exist.[100] He answered both questions in the negative, through a complex and original elaboration of the concept of norm. Canguilhem rejected the positivist concept of norm as average. Norms for him cannot be observed, or calculated from a collection of data, because they represent desired states of affairs. For him a norm is the expression of the values that human

94 Ibid., ch. 3, 'Claude Bernard et la pathologie expérimentale'.

95 Ibid., pp. 34–5.

96 Ibid., p. 61.

97 For a discussion of Canguilhem's attack on positivist or scientistic approaches in his body of work, see François Dagognet, 'Une œuvre en trois temps', *Revue de métaphysique et de morale* 90 (1985): 29–38.

98 Canguilhem, *Le normal et le pathologique*, p. 70.

99 Ibid., p. 13.

100 The two main parts of *Essai sur quelques problèmes concernant le normal et le pathological* (first part of *Le normal et la pathologique*) are entitled 'Is the pathological state merely a quantitative modification of the normal state?', and 'Do sciences of the normal and the pathological exist?', respectively.

beings spontaneously employ to organize their environment, and therefore cannot be reduced to an objective concept that can be determined scientifically. In his view, human beings are not the expression of norms, or the embodiment of given biological norms, as in the concept of the 'normal man'. Rather, they are normative in the sense that they establish norms.

In order to illustrate what Canguilhem argued, the example of the concept of 'good health' is fitting. 'Good health' is a norm by which we judge our health. However, it does not correspond to any average or general state of affairs. Human beings are not predominantly, or arguably ever, in the normative state of 'good health'. They commonly aim at being in 'good health', but this does not mean that they try to have average health, for instance trying to catch an average number of infections, but rather to get as close as possible to their normative concept of good health. This norm, however, cannot be determined once and for all, as for Canguilhem it is neither stable throughout time nor applicable to all individuals, but is rather dependent on many factors.[101] Some of them are rather obvious, like age: the norm of good health of a 20-year-old is not the same as that of a 90-year-old. Similarly, the level of health expected by a twenty-first-century individual living in a wealthy country is not the same as that of someone who lived in the Middle Ages. This is partly because of the obvious differences in health care and physical environment; the modern individual of course has by and large much higher expectations in terms of life expectancy, or recovery from accidents or illness. However, for Canguilhem the norm of good health also depends on cultural values. In an article on medicine, he pointed out that modern western individuals, who live in 'industrial and democratic' societies, even if they are Christian, would not share Pascal's view that illness is human beings' natural state, while health is a danger to their souls.[102] He also cited the example of yogis, who are able to alter their vegetative functions in a way that would be dangerously pathological for other people.[103]

For Canguilhem human beings creatively establish norms as a way to adapt to the environment and to circumstances. The few examples above show that cultural and social circumstances demand different responses from individuals. Individual circumstances are also different even within the same society or group, and each individual has to create his or her own norms and indeed change them in the course of her or his life.[104] The very process of falling ill and recovering is interpreted by Canguilhem as a process of variation of norms. Illness for Canguilhem is a type of normality in the sense that it has its own norms. If an individual contracts diabetes, she would have to establish new norms which are created by her new relationship with the environment.[105] The environment for Canguilhem is not more stable than

101 On this issue, see Guillaime Le Blanc, *La vie humaine: anthropologie et biologie chez Georges Canguilhem* (Paris: Presses Universitaires de France, 2002), ch. 2.

102 Canguilhem, 'Puissance et limites de la rationalité en médicine [1978], in Canguilhem, *Etudes d'histoire et de philosophie des sciences concernant les vivants et la vie*, p. 410.

103 Canguilhem, *Le normal et le pathologique*, pp. 106–107.

104 For an illuminating discussion of the concept of individuality in Canguilhem, see Gayon, 'The Concept of Individuality in Canguilhem's Philosophy of Biology'.

105 See Debru, *Georges Canguilhem, science et non-science*, pp. 35–7.

the life of individuals or societies. The environment in which individuals live for him has little to do with nature as described by science, which is regulated by laws. In his words:

> The living creature does not live among laws but among creatures and events which vary these laws. What hold up the bird is the branch and not the laws of elasticity. If we reduce the branch to the laws of elasticity, we must no longer speak of a bird, but of colloidal solutions. At such a level of analytical abstraction, it is no longer a question of environment for a living being, nor of health nor of disease. Similarly, what the fox eats is the hen's egg and not the chemistry of albuminoids or the laws of embryology. Because the qualified living being lives in a world of qualified objects, he lives in a world of possible accidents. Nothing happens by chance, everything happens in the form of events. Here is how the environment is inconstant. Its inconstancy is simply its becoming, its history.[106]

Canguilhem defined health as 'a margin of tolerance for the inconstancies of the environment'.[107] In other words, the more an individual is capable of adapting to the changed circumstances, and therefore of establishing new norms, the healthier she is. By contrast, the pathological state is the loss of the ability to adapt to new situations. For Canguilhem, the healthy living being adapts to her environment creatively: she imposes her own norms, and therefore her own values, to her environment, and in so doing, she changes it and makes it her own. As a consequence, the living being and her environment cannot be judged normal in separation, but only in their relation.[108]

In *Le normal et le pathologique*, and in later articles, Canguilhem directly discussed the question of the origin of the norm, whereas in his history of reflex he had simply received his norm from current science. The origin of the norm is clearly stated in his definition of normative:

> Normative, in philosophy, means every judgement which evaluates or qualifies a fact in relation to a norm, but this mode of judgement is essentially subordinate to that which established norms. Normative, in the fullest sense of the word, is that which established norms.[109]

For Canguilhem, the living being creates values which establish norms.[110] In the disciplines that Canguilhem analysed, especially medicine and psychiatry, the

---

106 Canguilhem, *Le normal et le pathologique*, pp. 130–31; the English translation is from Georges Canguilhem, *The Normal and the Pathological* (trans. Carolyn R. Fawcett; New York: Zone Books, 1989), pp. 197–8.

107 Canguilhem, *Le normal et le pathologique*, p. 120.

108 Canguilhem criticized psychology, and especially behaviourism and Pavlovism, precisely because they conceive of individuals as reacting to an environment on which they have little input or control. Jean-François Braunstein has discussed the evolution of Canguilhem's reflection on psychology in Jean-François Braunstein, 'Psychologie et milieu. Ethique et histoire des sciences chez Georges Canguilhem', in Braunstein (ed.), *Canguilhem: Histoire des sciences et politique du vivant*, pp. 63–89.

109 Canguilhem, *Le normal et le pathologique*, p. 77; the English translation is from Canguilhem, *The Normal and the Pathological*, pp. 126–7.

110 See Michel Fichant, 'Georges Canguilhem et l'idée de la philosophie', in Etienne Balibar et al. (eds), *Georges Canguilhem: Philosophe, historien des sciences. Actes du*

object is the living being, but the living being is also a subject. As he put it, ill individuals are subjects who regard themselves as such and regard their pain, and their representations of it, their angst, hopes and dreams, as their own individual possession.[111]

In medicine, as Canguilhem saw it, the 'objects' of knowledge create values and norms, and ultimately create themselves, in a way that the objects of mathematics and physics do not. Moreover, the objects of medicine are individuals, and their norms are set by each of them; in Canguilhem's words: 'the norm of pathology is an individual norm'.[112] It is the individual who decides whether she is sick or recovered, and whether she is 'normal'. This does not mean that the individual creates her norms independently, but rather within a biological and social system of values.[113] Nevertheless, medicine deals with individuals and with their environment, which does not show the regularities of the controlled environment of a physics experiment, as discussed above. In fact, all life sciences face for Canguilhem a similar problem; he argued that 'the science of life' has life both as object of study, and as subject, for it is the 'enterprise of living men'.

As Canguilhem described it, medicine is very different from physics and chemistry. Whereas physics and chemistry seek regularities, medicine deals with the individual; while their objects are clearly defined and constant, the objects of medicine are subjects who are all different from one another. Can medicine still be called a science? Canguilhem's answers to this question appear to be angled differently over the years, but they all suggest that medicine is more than a science. In his 1943 *Essai*, he described physiology, insofar as it is aimed at finding the regularities in the phenomena of life, as a science. However, when physiologists, as living beings, study the living, they practise a discipline that is 'more' and 'not less' than a science. Medical practice, in which, as Canguilhem put it, one has to face the setbacks of life, in other words the pathological, goes beyond scientific practice. He argued that the categories of health and illness are not 'biologically scientific and objective' but rather 'biologically technical and subjective'. Canguilhem here pointed out that medicine is first of all an intervention, and it is its 'technical' part that has the primacy over theory. Moreover, he stressed that subjectivity is not only impossible to eliminate from medicine, but it should not be eliminated.[114] He also defined medicine as an 'art of living' and as a 'technique or art at the crossroad of

---

colloque (6–7–8 décembre 1990) (Paris: Albin Michel, 1993), pp. 37–48.

111 Canguilhem, 'Puissance et limites de la rationalité en médicine' [1978], in Canguilhem, *Etudes d'histoire et de philosophie des sciences concernant les vivants et la vie*, p. 409.

112 Canguilhem, *Le normal et le pathologique*, p. 72.

113 In articles written late in his life, after Michel Foucault's *Birth of the Clinic: An Archaeology of Medical Perception* (trans. A. M. Sheridan; London: Routledge, 2003 [1963]), Canguilhem also discussed the role played by social 'surveillance' and by policies aimed at the improvement of the workers' health, see for instance: 'Les maladies' [1989], in Georges Canguilhem, *Ecrits sur la médicine* (Paris: Seuil, 2002).

114 Canguilhem, *Le normal et le pathologique*, pp. 149–50. On this issue, see Claude Debru, 'L'engagement philosophique dans le champ de la médicine: Georges Canguilhem aujourd'hui', in Braunstein (ed.), *Canguilhem: Histoire des sciences et politique du vivant*, pp. 45–62.

several sciences rather than, strictly speaking, one science'.[115] In his later writings, Canguilhem presented medicine as an 'evolutive sum of applied sciences'.[116] Here, with the adjective 'evolutive', Canguilhem stressed medicine's historical character, which pervades its methods and foundations. From this point of view, medicine is not different from any other science. He also emphasized the 'applied' character of medicine, just as he had stressed its technical character in *Le normal et le pathologique*. He did not just state the obvious, but, against positivistic hierarchies *à la* Comte, he also rejected the supposed superiority of the theoretical sciences over 'applied' sciences. Crucially, he pointed out that medicine is not an application of pre-existent theories, for its 'therapeutic project' is as original as its theoretical one. In other words, the technical side of medicine is not secondary to the theoretical part; in fact, Canguilhem emphasized again and again that medicine only exists because human beings experience pain and limitations to their life, and they want to restore a good relationship with their environment. He also acknowledged that medicine is a 'sum', indeed a synthesis, of different types of knowledge: technical knowledge, different sciences (e.g., chemistry and biology), economics and social sciences.

Although his claim is not as strong as Bachelard's, according to which science 'creates philosophy',[117] he nonetheless conferred to medicine a similar role as Brunschvicg did to mathematics and Bachelard to physics and chemistry: that of guiding philosophical reflection. His aim in *Le normal et le pathologique* was to 'integrate some of the methods and attainments of medicine into philosophical speculation'.[118] However, in Canguilhem's description, medicine does appear different from mathematics, physics and chemistry. As a consequence, what he contributed to philosophy by reflecting on medicine differs from what Brunschvicg and Bachelard contributed to it by focusing on mathematical and physical sciences. Although, as I have shown, these three projects stood in a relationship of continuity in many respects, Canguilhem's choice of focus profoundly changed the project of the history of the mind. Not only, like Bachelard, did he give great importance to the technical aspect of the discipline of his choice, but also shifted his focus away from the 'mind' – and in particular from the rational part of the mind – in two somewhat opposite ways. On the one hand, as Nikolas Rose has rightly pointed out, his history of reason is 'heterogeneous, regional, plural'.[119] His focus on concepts went further than Bachelard's 'regional rationalism'[120] in breaking up the object of which he wrote a history. On the other hand, the object of medicine cannot be reduced to the scheme of rationality. First of all, medicine, as Canguilhem saw it, is very much a

115 Ibid., pp. 77, 7.

116 Canguilhem, 'Le statut épistémologique de la médicine', in Canguilhem, *Etudes d'histoire et de philosophie des sciences concernant les vivants et la vie*, p. 423.

117 Bachelard, *Le nouvel esprit scientifique*, p. 7.

118 Canguilhem, *Le normal et le pathologique*, p. 8; the English translation is from: Canguilhem, *The Normal and the Pathological*, p. 34

119 Nikolas Rose, 'Life, Reason and History: Reading Georges Canguilhem', *Economy and Society* 27/2–3 (1998): 154–70 (159).

120 For Bachelard's regional rationalism, see Bachelard, *Le rationalisme appliqué*, ch. 7.

discipline that deals with bodies and minds as wholes.[121] Indeed, the environment in which minds and bodies live is just as integral to the object of medicine as they are. Whereas Bachelard's scientific object was rationally 'rectified', pure and regular, the objects of medicine, as Canguilhem present them, cannot be reconstructed rationally, in fact they are irregular and individual. Bachelard could separate scientific, that is rational, objects, from the emotional objects of our imagination; in Canguilhem's analysis of medicine, the 'angst and hopes' of the patient are as important parts of the object of knowledge as quantitave information, from which they cannot be easily separated. Qualitative knowledge for him is as an essential part of medicine as quantitative knowledge is.

He realized a philosophical version of Henri Berr's dream of universal history. Canguilhem's own history was far from all-encompassing, as I have discussed, but the object of his philosophical reflection was. His presentation of medicine did not allow the separation of reason, imagination, bodies, natural and social environments, economic circumstances and political power. For him, all is relevant and all is integrated. Although the roots of his project were in the philosophies of the Third Republic, his philosophical proposal went beyond his predecessors' dreams of a scientific rationality that alone could drive social and ethical change. Although Canguilhem was certainly a rationalist, he acknowledged a role to 'life' that could not be fully controlled, ordered and guided by rationality, and that could not be expelled from science.[122]

121 It is not surprising that some reflections of somatization in medicine have engaged with Canguilhem's philosophy: see Monica Greco, 'Between Social and Organic Norms: Reading Canguilhem and "Somatization" ', *Economy and Society* 27/2–3 (1998), pp. 234–48.

122 For this reason, Paul Rabinow, rightly called Canguilhem a 'vital rationalist': Rabinow, 'Introduction: A Vital Rationalist'.

Conclusion

Philosophical Questions across Fluid Disciplinary Boundaries

For Bachelard and Canguilhem science was essentially historical; indeed their philosophies of science were based on a study of science as a historical process. Their question was never how to interpret science once and for all, as this would have been contradictory, but rather how to evaluate its changes, either as overall ruptures in worldviews, methods and aims or as partial variations, in which concepts may survive theoretical revolutions. Their approaches have become synonymous with French philosophy of science as an autonomous tradition. It is their treatment of science as essentially historical that has been seen as the most fundamental difference with logical positivism in particular, and with large part of Anglo-American philosophy of science.[1] Institutionally, the connection of the history and the philosophy of science in France is well illustrated by the Sorbonne chair that first Rey, and then successively Bachelard and Canguilhem occupied: that of History and Philosophy of the Sciences. The institute that Rey founded, and the other two directed, the Institut d'histoire des sciences et techniques, did not even have the term 'philosophy' in its name. However, Rey's guiding idea, which shaped syllabi and research activities, was to connect closely history and philosophy, and to evaluate the history of science philosophically. The following two directors, if anything, had a more philosophical approach to the history of science than Rey did.

History was crucial in Bachelard's and Canguilhem's philosophies, and certainly marked a fundamental difference with other approaches to science. However, the importance of history was neither specific to their philosophies, nor to the philosophy of science. In fact, the important place that the history of philosophy came to acquire in the first part of the twentieth century made the connection between history and philosophy not only present in the work of many scholars, but widely discussed and reflected upon. It was by no means the case that all French philosophy was historical; many doctrines were not, and there were even those who, like Jean Hyppolite, lamented that French philosophy 'from Descartes to Bergson' had rejected history.[2] However, as Léon Brunschvicg remarked, and the analysis of professorships, courses, doctoral dissertations and publications show, the history of philosophy was particularly important at the Sorbonne. Needless to say, what was important at the Sorbonne could not fail to play an extremely central role in the national academic life. More specifically, it was in that institution that Bachelard elaborated his philosophy, first as a student of Léon Brunschvicg, and later in life as a professor. Indeed, the former developed many of the central questions and themes of his teacher, who held the chair of History of Modern Philosophy. It should not be forgotten that Brunschvicg

1 See Gutting, 'French philosophy of science'. Gutting's view is widely held and indeed well-founded.

2 Hyppolite, *Introduction à la philosophie de l'histoire de Hegel*, p. 123.

was a historian of philosophy, and that his reflection on past doctrines, and above all on the development of ideas and philosophical questions in history was central to his own approach, and indeed to subsequent doctrines, including Bachelard's.

Morever, it is not only the importance of history that characterized Bachelard's and Canguilhem's philosophies, but also a set of philosophical questions that they inherited from research programmes developed during the Third Republic. These research programmes, which were aimed at studying the mind '*a posteriori*', by using history as its 'laboratory', did not specifically belong to the philosophy of science. Among historians of philosophy, Brunschvicg himself was the most prominent scholar to embark upon this type of project. His colleague and predecessor in the chair of History of Modern Philosophy, Lévy-Bruhl, aimed at answering very similar questions, but rather than history he turned to ethnological reports. However, the fundamental aims of their projects were similar: to analyse the different ways in which the mind works, and understand different worldviews and mental categories. Indeed, Lévy-Bruhl regarded his analysis of 'primitive' mentality as just another way of answering philosophical questions. Although his attention might have been focused on the mentalities of peoples who were his contemporaries, but lived in very different types of societies from his own, his doctrine had a great impact on French history and philosophy of science.

Brunschvicg's and Lévy-Bruhl's approaches, however, consciously differed from those that were more mainstream in the history of philosophy. They read philosophical, scientific or ethnological texts in order to find out the mentalities that had produced them. Just as Lévy-Bruhl did not aim to assess whether what 'primitives' said about nature was true or false, but rather to understand what mental mechanism and worldviews made their assertions possible, so Brunschvicg – at least with his 'historian of the mind' hat on – did not aim to engage in a direct dialogue with Kant, but to establish how he could judge those particular mental categories as 'absolute'. Brunschvicg aimed to show that Kant's philosophy was the result of a particular moment in intellectual history, and that it could not have come about without the Newtonian worldview. In this manner, Brunschvicg read philosophical texts as relative to a particular historical moment, and in so doing challenged the intrinsic timeless value that philosophers attached to them. In fact, for Lévy-Bruhl and Brunschvicg's purposes, it was not necessary to analyse philosophical texts in particular. This is why Lévy-Bruhl could turn to ethnological reports, and the latter could examine both philosophical and scientific works.

Scholars who aimed to study the mind in its diverse approaches to reality, not only turned to different objects of study, but also employed different methods. Indeed, these projects were not only limited to the philosophy of science and the history of philosophy, but developed across several disciplines, including the social sciences and the history of science. However, it would be incorrect to interpret these research programmes simply as interdisciplinary. It was not the case that these disciplines existed in complete autonomy, and happened to share particular questions. Emergent disciplines, notably sociology, aimed to pursue a more scientific approach to the study of the mind than philosophy had traditionally offered. Intellectually and institutionally linked to philosophy, emergent disciplines often aimed to mark their differences with philosophy. Nevertheless, philosophers, sociologists, psychologists, historians of religions, historians of science and other scholars debated with each other in places such as the Société française de philosophie and the international conferences of philosophy. The *Revue de métaphysique et de morale* and the *Revue philosophique* published articles by Brunschvicg and Lalande along those by Mauss, Dumas and Davy. The institutional and intellectual links among the practitioners of disciplines that still were not as separate as we might imagine them today, made the discussions rich and challenging. Brunschvicg might have been Professor of History of Modern Philosophy, but his references to the sociologist Durkheim, and to his colleague Lévy-Bruhl's studies on primitive mentalities were utterly unsurprising. In fact, it would have been remarkable if he had failed to mention their works when discussing topics that they had examined. The wide range of references found in the works of many French scholars was not just an accident of disciplinary history, but indeed the result of their comprehensive education, and indeed of the value that they themselves often attached to the ideal of *culture générale* that shaped school syllabi.

It is therefore artificial to focus on the history of a particular discipline in order to explain the background of historical epistemology, and so is to reconstruct twentieth-century French 'philosophy of science' by excluding scholars who were highly relevant to philosophical reflection on science, but who might not even have had a scholarly interest in science, as in the case of Lévy-Bruhl. Moreover, the very fact that much of French philosophy of science has analysed the history of science suggests that historians of science, and indeed other historians, were an important part of their intellectual milieu. This was indeed the case, and the distinction between philosophers and historians was not always clear, and sometimes not even helpful, if imagined as an insurmountable boundary. Scholars such as Brunschvicg, Rey, Febvre, Metzger, Koyré and Bachelard all aimed to answer questions about the mind and about the mutations in ways of reasoning and representing the world. They had many chances to discuss how to confront these questions with one another, including in seminars and by reviewing one other's books. Their schooling and careers interlinked in many ways: some had attended or taught in the same *lycée*, or they were together at the Ecole normale (and in both places crucial alliances were formed), and at the Sorbonne as students. Those who became Sorbonne professors shared students (who included the younger scholars in this cohort, such as Bachelard) among themselves, but also with Metzger and Koyré, as they lectured at Rey's Institut d'histoire des sciences et techniques.

However, their close links and shared philosophical questions did not exclude differences in methods and priorities. Although they all analysed history and they all asked philosophical questions about the mind, some of them regarded themselves predominantly as historians and other as philosophers. These differences did not prevent their interactions. Even Febvre, who is not among the scholars on whom I have directly focused, as he regarded himself as a 'pure' historian and was openly opposed to the philosophers' approach, after all closely collaborated with Rey, who in turn was seen as a philosopher by the 'historian' Metzger. To close the circle, Febvre perceived Metzger's talks on historiography as philosophical, and therefore, in his view, not really relevant to history writing.

These scholars' different disciplinary choices become clearer and acquire a richer meaning precisely when considered in relation to one another, as they themselves did. For instance, Rey wanted to create a space for a type of history of science that, on the one hand, would be more historically accurate than that practised by philosophers such as Brunschvicg and Meyerson, but that, on the other hand, would not be separated from the history of philosophy, and would aim to answer philosophical questions, unlike the 'pure' history of science practised by Pierre Boutroux at the Collège de France.

Social and Disciplinary Marginality and Personal Strategies

French academia greatly expanded in the first decades of the twentieth century, both in terms of numbers of students and staff, and in terms of disciplines. Beyond quantitative assessments, this expansion facilitated a certain degree of social mobility, which appeared to be greater than in later periods, including the 1960s. Individuals from non-intellectual backgrounds, who, in other words, were not 'inheritors' in Bourdieu's sense, appeared to obtain posts of importance in larger numbers, and, contrary to a widely held perception, represented almost two-thirds of the professors appointed at the Paris Faculty of Letters in the period 1900–1939. Moreover, the increased presence of academics from a Jewish background, who previously had been all but excluded from academia, also represented a novelty. The academics whose fathers had different occupations from themselves, had diverse backgrounds, as some were from wealthy families, while a smaller number of them were from lower social strata. Lévy-Bruhl and Brunschvicg belonged to the latter group. Indeed, in the context of their fellow historians of philosophy, they stood out as they, both Jewish, also had the weakest social and inherited economic capital.

The intellectual and academic success that they enjoyed, along with other intellectuals who did not have an 'orthodox' social profile, has often hidden the fact that they were in a minority. Lévy-Bruhl and Brunschvicg attached great importance to the secular republican institutions which had enabled them to rise from the humble and marginal conditions of their parents to recognition, fame and academic power. Moreover, their discipline was philosophy, which was a core subject in French education, both at secondary and tertiary level. Their speciality, the history of philosophy, combined the mainstream appeal of philosophy with the prestige of tradition and of philological studies.

The majority of historians of philosophy appointed at the Sorbonne and the Collège de France between the beginning of the twentieth century and the end of the inter-war period did not display either the social profile or the secular and political progressive ideas that have often been attributed to French professors in this period. A simple majority of Brunschvicg and Lévy-Bruhl's fellow historians of philosophy were from Catholic families, and for some of them, notably Etienne Gilson, religion was an important part of their scholarly and professional life. The cause of secularism was to be defended not only in society but indeed in their academic work as Brunschvicg did against Gilson. The innovation in traditional disciplines such as French and philosophy that progressive scholars, including Lévy-Bruhl and

Brunschvicg, aimed to implement were under attack by conservatives and far-right groups. Indeed, one of their own fellow Sorbonne historians of philosophy, Rivaud, had strong links with the monarchist and anti-Dreyfusard Action française.

Although Lévy-Bruhl's and Brunschvicg's political ideas were not the same, for the former was a socialist, while the latter was closer to the more moderate Radical party, for both of them secularism and republicanism were important not only in their lives but also in their work. They did not primarily aim at preserving and keeping alive an intellectual tradition, unlike many historians of philosophy. Both of them studied earlier ideas, and, in the case of Lévy-Bruhl, ideas from other cultures, in order to distinguish modern, scientific and secular ways of thinking from 'primitive', magical and religious approaches to reality. In their different ways, they counterpoised modernity and tradition, science and religion. Their republican ideals were also the ideals that, if put into practice, would enable persons like them to occupy important positions in French institutions. Their cultural battles helped to create and consolidate spaces not only for academics from a wider range of backgrounds, but also for new ways of conceiving of intellectual history: Lévy-Bruhl contributed to the creation of a new discipline, ethnology, and had a great intellectual impact on other new disciplines, including the history of science; Brunschvicg proposed a new approach to science, which created the basis for historical epistemology.

Whereas philosophy in the first decades of the twentieth century was a leading discipline in French education, the history of science was still struggling to find a stable space in academia. Abel Rey, who acquired his cultural capital in philosophy, and invested it the history of science, created it at the Sorbonne. His own Sorbonne chair of History and Philosophy of the Sciences was still a philosophy chair. Rey, who was also from an non-intellectual family and a Dreyfusard, developed his approach to the history of science in close contact with the intellectuals of the Centre de synthèse, and published his books on ancient science in Henri Berr's series. For a marginal discipline such as the history of science, extra-academic spaces, and the Centre de synthèse in particular, played a crucial role. Scholars at the margins of, or excluded from, French universities, such as Metzger and Mieli, were instrumental in creating national and international networks of historians of science and in promoting this discipline at all levels. Henri Berr, from a non-intellectual family, never managed to obtain a chair in a higher-education institution, but his journal, his Centre and his many initiatives created important spaces for innovative intellectual enterprises, and indeed for a new discipline such as the history of science, that had a limited space in universities. The Centre created a crucial space for intellectual exchange, including those scholars who did not share the positivistic views of its director. Metzger is a case in point: she might have disagreed with Berr about history writing, but still shared the progressive politics of the Centre de synthèse, which gave her a sought-after role of responsibility, at a time when the republican institutions were clearly not ready to admit a brilliant intellectual, whose gender and education differed from the norm.

Alexandre Koyré, whom Metzger introduced to the Centre, did not establish a long-term collaboration with the scholars there. The opposition of Aldo Mieli was certainly the main problem, but Koyré also did not seem to fit well in a milieu in which scholarly pursuits went hand-in-hand with the active promotion of progressive political ideas and of secularism. Indeed, his own account of the history of the mind contradicted the opposition between scientific and religious thought that many at the Centre favoured. Koyré became an internationally famous historian of science, but in France did not lead the discipline. Although, in the 1950s, he presented himself as the heir of the French tradition that combined philosophy and the history of science, he was not really seen as an 'inheritor' in the broader sense of the term. Indeed, notwithstanding his claims, the Collège de France chair for which he was applying went to Martial Gueroult, a historian of philosophy who was the heir of the mainstream tradition from which Lévy-Bruhl and Brunschvicg had departed. It was Gueroult who claimed that Brunschvicg's history of the mind in effect expelled the history of philosophy from philosophy; by contrast, he aimed to practise the history of philosophy as a philosophical discipline.

Koyré had moved from the history of philosophical and religious ideas to the history of science as the history of the mind. He created his own Centre de recherche d'histoire des sciences et des techniques at the Ecole pratique des hautes études, but neither the Collège de France nor the Sorbonne were open to him. The 'heir' at the Sorbonne was Gaston Bachelard, student of Rey, who had created the space for philosophical history of science, and Brunschvicg, who had made the project of writing the history of the mind through the history of science a highly respectable enterprise.

The Writers of History and the Objects of Knowledge

There is no doubt that such scholars as Lévy-Bruhl, Brunschvicg, Berr, Rey and Metzger, identified modernity with scientific rationality, progress, self-reflectivity and, as Rey put it, with progressive 'liberation' from myth, religion and dogma. Despite their philosophical differences (for instance their different attitude towards positivism), they shared an Enlightenment project, in which truths proceed from rationality rather than tradition and religion. In their view, rational knowledge is objective, can in principle be acquired by anyone, and is independent of social privileges and personal desires. As they saw it, objective knowledge was not the reserve of a particular class of rational beings. Indeed, Lévy-Bruhl argued that it did not depend on the knower's ethnicity, as he pointed out that 'primitive' children, when taught, learned science just like their European peers. Many of these scholars were Jewish, and were proving with their intellectual achievements that religious background could not justify the exclusion from French universities, which right-wing activists appeared to seek. For some of them it did not depend on gender, either: as Metzger put it to Meyerson 'in the Republic of the mind, we are all equals'.[3]

However, the set of projects of writing the history of the mind that I have discussed were based on the analysis of differences: differences between the modern and primitive, between science and metaphysics, between rational inquiry and imaginative theories, and also between current scientific rationality and its previous

3 See Chapter 5. Brunschvicg's involvement with the Ligue d'electeurs pour le suffrage des femmes and Lévy-Bruhl's support of Metzger's scholarly activities are also worth mentioning.

historical stages. Scholars who pursued these projects did not have identical views of these oppositions. Lévy-Bruhl proposed a binary opposition between modern and primitive mentalities, although at various stages in the development of his own doctrine, he admitted the possibility of the co-existence of modern and primitive thought in the same individual. Brunschvicg interpreted Lévy-Bruhl's representation of primitive mentality as that of the first stage of human beings' intellectual progress towards a fully rational, scientific, self-reflective and secular approach to reality. He focused on all the intermediary stages that reason had gone through in its history. For him it was not only important to analyse the difference between primitive and religious thought on the one hand, and modern rationality on the other, but also the history of rational thought itself, and its potentiality for future development, as well as the threats to it from irrational thought, as exemplified by the rise of Nazism. In his presentation of the *outillage mental*, Rey discussed both the history of reason from primitive mentality to the present, and various ways of thinking as synchronically distinct from one another. His advocacy of 'western thought' was at least as uncompromising as Brunschvicg's apology for 'European thought'; he presented it as opposed not only to primitive thought, but also to eastern mentality, which, according to him, had developed along primitive lines, and even to eastern European mentality which in his view had retained some primitive traits.

That science, and current rational inquiry, is the norm by which all other activities should be judged was assumed by these Third Republic scholars, especially the more philosophically oriented ones. For them, those who possessed rational and self-reflective knowledge become both judges of all other producers of knowledge (or, from the point of view of the former in particular, the producers of false beliefs), and of themselves. Their double role enabled them on the one hand to judge 'primitive' people and 'eastern' people, and on the other to judge past and present developments of rational knowledge. Individual scholars had different views of what fell into the 'primitive' and in the 'rational' categories. For instance, Brunschvicg presented the history of the (European) mind from Thales to Einstein as the progress of rationality and the retreat of dogmas, notwithstanding the setbacks that the mind experienced. For him, Socrates and Spinoza belong to the same history of reason as the theory of relativity. By contrast, Bachelard placed all knowledge prior to the nineteenth century in the category of 'pre-scientific', or primitive, thus shortening the history of the scientific mind by well over two millennia. On the other hand, Bachelard re-admitted the primitive within the modern individual, as he believed that the primitive mind lived on in modern human beings and was responsible for poetry and art. However, even for him, it is the rational part that judges, and watches over, the imaginative part of the mind. Although they believed that in principle anybody could be a protagonist in the advancement of rational knowledge, in practice science had developed in Europe and then in the 'west'. In their eyes, it followed that modernity, progress and indeed leadership did belong to that part of the world.

The more historically oriented scholars, such as Metzger, made a point of understanding rather than judging ways of thinking that differed from current ones. Lévy-Bruhl also aimed to show that the beliefs of 'primitive' people should not be judged according to the worldviews and logic of modern science. However, even for these intellectuals the subject of knowledge is the modern, and western, scholar who is rational and self-reflective. Self-reflectivity was for all these scholars a crucial characteristic of scientific knowledge: for Brunschvicg and Bachelard self-reflectivity is at the very core of rational inquiry, which consistently for them is always discursive rather than intuitive. Even Metzger, who thought that 'spontaneous' thought (including religion) played a role in scientific knowledge, distinguished the latter from the former on the basis of the fact that scientists critically evaluate their own knowledge and activities. Metzger's alchemist who relies on 'spontaneous' thought and Lévy-Bruhl's primitives are not objects of their own knowledge, but rather of the modern and western scholar's research.

The consequence of their perspectives is a separation between rational scholars who objectify both their own knowledge and others' beliefs on the one hand, and those individuals whose minds are objects of the rational scholar's knowledge, but not of their own. The rational and self-reflective knowledge that philosophers such as Brunschvicg and Bachelard described was not the general possession of all modern individuals, but of a restricted number of them. Bachelard's 'epistemological rupture' between 'common knowledge' and 'scientific knowledge' (which reflected a common distinction in French philosophy) excluded the holder of the former from being a subject of authentic knowledge. Both he and Brunschvicg thought that the theory of relativity had revolutionized our way of thinking and started a new age in the history of the mind. However, the dramatic intellectual changes that they described arguably affected a very small number of people.

Scholars such as Lévy-Bruhl, Berr, Brunschvicg, Rey and Brunschvicg regarded scientific rationality as the mark of modernity and, in different ways, as affording the overcoming of exclusions and restrictions that tradition and religion had created. Their own doctrines, however, seem to recreate exclusions, this time based on knowledge. The interest in education that most of them displayed, and Bachelard's advocacy of life-long education, and of a society 'made for school', were ways of promoting inclusion and spreading knowledge to everybody.[4] The tension between their aspiration to universality and the identification of universality with Europe, the 'west', or western science, is not specific to these philosophers and historians of science, and is a problem that goes far beyond the scope of the present book. However, within the set of projects of writing the history of the mind, Canguilhem's proposal appears to redress the balance, at least partially, between the subjects and objects of knowledge. His view of medicine and psychiatry was based on the full appreciation that their objects of knowledge are also subjects, whose judgements, hopes and angst are the motor of the development of knowledge. Consistently, he argued that the rationality of scientists and technicians depends on the rationality of those who are directly affected by the advances in science. The creation of norms for Canguilhem is not the reserve of experts but it is rather a complex process in which every aspect of life – biological, social, cultural, political and institutional – contribute.

4 See Bachelard, *La formation de l'esprit scientifique: contribution à une psychanalyse de la connaissance objective*, p. 252.

Bibliography

Manuscript sources

Archives Nationales, Paris	AJ/16
Houghton Library	bMS Am 1803/1803.1
Wellcome Institute Library	PP/CJS
Yale University Library	Sigerist Papers (788)

Other references

Académie Internationale d'Histoire des Sciences, 'Elections de 1950', *Archives Internationales d'Histoire des Sciences* 12 (1950): 657.

—— 'Election du Secrétaire perpétuel', *Archives Internationales d'Histoire des Sciences* 33 (1955): 436.

Agathon, *L'esprit de la nouvelle Sorbonne. La crise de la culture classique. La crise du français* (Paris: Mercure de France, 1911 [1910]).

Althusser, Louis, 'Pierre Macherey, "La philosophie de la science de Georges Canguilhem": Présentation', *La Pensée* 133 (1964): 50–54.

Ambrière, Francis, 'Société des amis de l'histoire des sciences', *Thalès* 3 (1936): 249–61.

Amsterdamski, Stefan, 'La philosophie en science', *History and Technology* 4 (1987): 103–13.

Anonymous, 'Chronique', *Annales de l'Université de Paris* 5/3 (1930): 417–21.

—— 'Emile Durkheim (1858–1917)', *Revue de métaphysique et de morale* 24 (1917): 749–51.

—— 'Frédéric Rauh (1848–1909)', *Revue de métaphysique et de morale* 17, supplément de Mars (1909): 2–3.

—— 'Programme détaillé du certificat d'histoire et philosophie des sciences et du diplôme de l'Institut', *Thalès* 3 (1936): 233–47.

Anonymous [Henri Berr], 'Sur notre programme', *Revue de synthèse historique* 1/1 (1900): 1–8.

Anonymous [E. Bréhier], 'Notre programme', *Revue d'histoire de la philosophie* 1/1 (1927): 1–4.

Antognazza, Maria Rosa, 'Leibniz and the Post-Copernican Universe. Koyré Revisited', *Studies in History and Philosophy of Science* 34 (2003): 309–27.

Anzieu, Didier, 'La psychanalyse au service de la psychologie', *Regards sur la psychanalyse en France. Nouvelle Revue de psychanalyse* 20 (1979): 59–75.

Ardigò, Roberto, *La morale dei positivisti* (Milano: Natale Battezzati, 1889).

—— *La psicologia come scienza positiva* (Mantova: Viviano Guastalla, 1870).

Aron, Raymond, *Mémoires* (Paris: Robert Laffont, 2003 [1983]).

Association des lauréats du Concours général website: http://concoursgeneral.free.fr (accessed 18 April 2007).

Babini, José, 'Para una bibliografia de Aldo Mieli', *Physis* 21 (1979): 357–424.

Bachelard, Gaston, *Essai sur la connaissance approchée* (Paris: Vrin, 1987 [1927]).

—— *Etude sur l'évolution d'un problème de physique: la propagation thermique dans les solides* (Paris: Vrin, 1973 [1927]).

—— *Etudes* (Paris: Vrin, 1970).

—— *The Formation of the Scientific Mind* (trans. Mary McAllester Jones; Manchester: Clinamen Press, 2002).

—— *Fragments d'une poétique du feu* (Paris: Presses Universitaires de France, 1988).

—— 'Idéalisme discursif', *Recherches philosophiques* 4 (1934–35): 21–9.

—— 'Institut d'histoire des sciences et techniques: rapport sur l'année scolaire 1940–41', *Annales de l'Université de Paris* 16 (1941): 325.

—— *L'activité rationaliste de la physique contemporaine* (Paris: Presses Universitaires de France, 1951).

—— *L'air et les songes. Essai sur l'imagination du mouvement* (Paris: Corti, 1972 [1943]).

—— *La formation de l'esprit scientifique: contribution à une psychanalyse de la connaissance objective* (Paris: Vrin, 1993 [1938]).

—— *La philosophie du non. Essai d'une philosophie du nouvel esprit scientifique* (Paris: Presses Universitaires de France, 1988 [1940]).

—— *La poétique de la rêverie* (Paris: Presses Universitaires de France, 1960).

—— *La poétique de l'espace* (Paris: Presses Universitaires de France, 1964 [1957]).

—— *La psychanalyse du feu* (Paris: Gallimard, 1949 [1938]).

—— *La terre et les rêveries de la volonté. Essai sur l'imagination de la matière* (Paris: Corti, 1992 [1947]).

—— *La terre et les rêveries du repos. Essai sur les images de l'intimité* (Paris: Corti, 1992 [1948]).

—— *La valeur inductive de la relativité* (Paris: Vrin, 1929).

—— *Lautréamont* (Paris: Corti, 1986 [1939]).

—— *L'eau et les rêves. Essai sur l'imagination de la matière* (Paris: Corti, 1942).

—— *Le matérialisme rationnel* (Paris: Presses Universitaires de France, 1972 [1953]).

—— 'Le monde comme caprice et miniature', *Recherches philosophiques* 3 (1933–34): 306–20.

—— *L'engagement rationaliste* (Paris: Presses Universitaires de France, 1972).

—— *Le nouvel esprit scientifique* (Paris: Presses universitaires de France, 1991 [1934]).

—— *Le pluralisme cohérent de la chimie moderne* (Paris: Vrin, 1973 [1932]).

—— *Le rationalisme appliqué* (Paris: Presses Universitaires de France, 1986 [1949]).

—— 'Logique et épistémologie', *Recherches philosophiques*, 6 (1936–37): 410–13.

—— 'Noumène et microphysique', *Recherches philosophiques* 1 (1931–32): 55–65.

—— *The Psychoanalysis of Fire* (trans. Alan C.M. Ross; Boston: Beacon Press, 1964).

Bachelard, Gaston et al., 'Sur la continuité et la multiplicité temporelles. Séance du 13 mars 1937', *Bulletin de la Société française de Philosophie* (1937): 53–81.

Backès-Clément, C. and B. Pingaud, 'Le dernier alchimiste', *L'Arc* 42 (1970): 1–3.

Barberis, Daniela S., 'Moral Education for the Elite of Democracy: The Classe de Philosophie between Sociology and Philosophy', *Journal of the History of the Behavioral Sciences* 38/4 (2002): 355–69.

Barbier, Maurice, *La laïcité* (Paris: L'Harmattan, 1995).

Belaval, Yvon, 'Les recherches philosophiques d'Alexandre Koyré', *Critique* 20/206 (1964): 675–704.

—— (ed.), *Histoire de la philosophie* (3 vols; Paris: Gallimard, 1973).

Belot, Gustave et al., 'La morale positive. Examen de quelques difficultés. Séance du 26 mars 1908', *Bulletin de la Société française de Philosophie* 8 (1908): 161–215.

Ben-David, J., 'Roles and Innovation in Medicine', *American Journal of Sociology* 65 (1960): 557–68.

Ben-David, J. and Randall Collins, 'Social Factors in the Origins of a New Science: The Case of Psychology', *American Sociological Review* 31 (1966): 451–65.

Bensaude-Vincent, Bernadette, 'Chemistry in the French Tradition of Philosophy of Science: Duhem, Meyerson, Metzger and Bachelard', *Studies in History and Philosophy of Science* 36 (2005): 627–48.

—— 'Hélène Metzger's *La Chimie*: A Popular Treatise', *History of Science* 25/1 (1987): 71–84.

—— *Langevin: science et vigilance* (Paris: Belin, 1987).

—— 'Présences scientifique aux semaines de synthèse', in Agnès Biard, Dominique Bourel and Eric Brian (eds), *Henri Berr et la culture du XXe siècle. Histoire, science et philosophie* (Paris: Albin Michel/Centre international de synthèse, 1997), pp. 219–30.

Bergson, Henri, 'La philosophie française (Tableau récapitulatif destiné à l'Exposition de San Francisco)', *Revue de Paris* 22/3 (1915): 236–56.

—— *Lettere a Xavier Léon e ad altri* (Naples: Bibliopolis, 1992).

—— *L'évolution créatrice* (Paris: Alcan, 1907).

—— 'L'intuition philosophique', *Revue de métaphysique et de morale* 19 (1911): 809–27.

—— 'L'intuition philosophique', in International Congress of Philosophy *Atti del IV Congresso Internazionale di filosofia Bologna MCMXI*, vol. I (Genova: Formiggini, 1912 [1911]), pp. 174–92.

Berr, Henri, 'Au bout de dix ans', *Revue de synthèse historique* 21 (1910): 1–13.

—— 'Au bout de trente ans. La Revue organe du Centre', *Revue de synthèse* 51 (1931): 3–8.

—— 'Henri Berr par lui-même', *Revue de synthèse* 85 (1964 [1942]): 1–4.

—— 'Introduction à une Histoire Universelle', *Revue de synthèse historique* (1920): 17–34.

—— *La synthèse en histoire; essai critique et théorique* (Paris: Alcan, 1911).

—— *L'avenir de la philosophie: Esquisse d'une synthèse de connaissances fondée sur l'histoire* (Paris: Hachette, 1899).

Berr, Henri and Lucien Febvre, 'History', in Edwin R.A. Seligman (ed.), *Encyclopaedia of the Social Sciences*, vol. 7 (New York: Macmillan, 1932), pp. 357–68.

Berthelot, Marcelin, *Collections des anciens alchimistes grecs* (2 vols; Paris: Steinheil, 1887).

—— *Introduction a l'étude de la chimie des anciens et du Moyen âge* (Paris: Steiheil, 1889).

—— *La chimie au Moyen âge, vol. I: essai sur la transmission de la science antique au Moyen âge* (Paris: Imprimerie nationale, 1893).

—— *La chimie au Moyen âge, vol. II: l'alchimie syriaque* (Paris: Imprimerie nationale, 1893).

—— *La chimie au Moyen âge, vol. III: la chimie arabe* (Paris: Imprimerie nationale, 1893).

—— *La révolution chimique; Lavoisier* (Paris: Alcan, 1890).

—— *Les origines de l'alchimie* (Paris: Steinheil, 1885).

Biagioli, Mario, 'Meyerson, Science and the Irrational', *Studies in History and Philosophy of Science* 19 (1988): 5–42.

Bing, François and Jean-François Braunstein, 'Entretien avec Georges Canguilhem', in François Bing, Jean-François Braunstein and Elisabeth Roudinesco (eds), *Actualité de Georges Canguilhem. Le normale et le pathologique* (Paris: Synthélabo, 1998), pp. 121–35.

Bloch, Marc, *La société féodale: la formation des liens de dépendance* (Paris: Albin Michel, 1939).

—— *La société féodale: les classes et le gouvernement des hommes* (Paris: Albin Michel, 1940).

—— *Les rois thaumaturges: étude sur le caractère surnaturel attribué à la puissance royale, particulièrement en France et en Angleterre* (Strasbourg: Faculté des Lettres de l'Université de Strasbourg, 1924).

Blondel, Maurice, 'Y a-t-il une philosophie chrétienne?', *Revue de métaphysique et de morale* 38 (1931): 599–606.

Bohr, Niels, 'Quantum Mechanics and Physical Reality', in John A. Wheeler and Wojciech H. Zurek (eds), *Quantum Theory and Measurement* (Princeton, New Jersey: Princeton University Press, 1983 [1935]), p. 144.

Boirel, René, *Brunschvicg: sa vie, son œuvre, avec un exposé de sa philosophie* (Paris: Presses Universitaires de France, 1964).

Bompaire-Evesque, Claire-Françoise, *Un débat sur l'université au temps de la Troisième République. La lutte contre la Nouvelle Sorbonne* (Paris: Aux amateurs de livres, 1988).

Bonicalzi, Francesca, *Leggere Bachelard: le ragioni del sapere* (Milan: Jaca Book, 2007).

Bouglé, Célestin, *Humanisme, sociologie, philosophie. Remarques sur la conception française de la culture générale* (Paris: Hermann, 1938).

—— *Les maîtres de la philosophie universitaire en France* (Paris: Maloine, 1938).

—— (ed.), *Encyclopédie française, vol. XV: Education et Instruction* (Paris: Société de gestion de l'Encyclopédie française, 1939).

Bouglé, Célestin et al., 'La sociologie du Proudhon', *Bulletin de la Société française de Philosophie* 12 (1912): 169–213.

Bourdieu, Pierre, *Distinction: A Social Critique of the Judgment of Taste* (trans. Richard Nice; London: Routledge, 1989 [1979]).

—— 'Georges Canguilhem: An Obituary Notice', *Economy and Society* 27/2–3 (1998): 190–92.

—— *Homo Academicus* (trans. Peter Collier; Cambridge: Polity Press, 1988 [1984]).

—— 'The Peculiar History of Scientific Reason', *Sociological Forum* 6/1 (1991): 3–26.

—— *The State Nobility* (trans. Lauretta C. Clough; Cambridge: Polity Press, 1998 [1989]).

Bourdieu, P. and Jean-Claude Passeron, *The Inheritors: French Students and their Relation to Culture* (trans. Richard Nice; Chicago: University of Chicago Press, 1979 [1964]).

—— *Reproduction in Education, Society and Culture* (trans. Richard Nice; London: Sage, 1977 [1970]).

Bourdieu, Pierre, Jean-Claude Chamboredon and Jean-Claude Passeron, *The Craft of Sociology: Epistemological Preliminaries* (trans. Richard Nice; Berlin: de Gruyter, 1991 [1968]).

Bourel, Dominique, 'Présentation: Henri Berr (1863–1954)', in Agnès Biard, Dominique Bourel and Eric Brian (eds), *Henri Berr et la culture du XX*[e] *siècle. Histoire, science et philosophie* (Paris: Albin Michel/Centre international de synthèse, 1996).

Bourgeois, Bernard, 'Lévy-Bruhl et Hegel', *Revue philosophique* 114 (1989): 449–51.

Boutroux, Emile, *De la contingence des lois de la nature* (Paris: Presses Universitaires de France, 1990 [1874]).

—— 'Histoire et synthèse', *Revue de synthèse historique* 1/1 (1900): 9–13.

—— *Pascal* (Paris: Hachette, 1900).

—— 'Rôle de l'histoire de la philosophie dans l'étude de la philosophie', in Edouard Claparède (ed.), *Congrès de Genève 1904* (Geneva: Künding, 1905), pp. 49–68.

Boutroux, Emile, et al, 'Science et religion. Séance du 19 novembre 1908', *Bulletin de la Société française de Philosophie* 9 (1908): 19–73.

Boutroux, Pierre, *L'idéal scientifique des mathématiciens, dans l'antiquité et dans les temps modernes* (Paris: Alcan, 1920).

Bouvier, Robert, 'Henri Berr et son œuvre', *Revue de synthèse* 85 (1964): 39–50.

Braunstein, Jean-François, 'Abel Rey et les débuts de l'Institut d'histoire des science et techniques (1932–1940)', in Michel Bitbol and Jean Gayon (eds), *L'épistémologie française, 1830–1970* (Paris: Presses Universitaires de France, 2006), pp. 173–91.

—— 'Canguilhem avant Canguilhem', *Revue d'histoire des sciences* 53/1 (2000): 9–26.

—— 'Psychologie et milieu. Ethique et histoire des sciences chez Georges Canguilhem', in Jean-François Braunstein (ed.), *Canguilhem: Histoire des sciences et politique du vivant* (Paris: Presses Universitaires de France, 2007), pp. 63–89.

—— (ed.), *Canguilhem: Histoire des sciences et politique du vivant* (Paris: Presses Universitaires de France, 2007), pp. 63–89.

Bréhier, Emile, 'Comment je comprends l'histoire de la philosophie', *Les Etudes philosophiques*, nouvelle série, 2 (1947): 105–14.

—— 'The Formation of our History of Philosophy', in Raymond Klibansky (ed.), *Philosophy and History. Essays presented to Ernst Cassirer* (Oxford: Claredon, 1936), pp. 159–72.

—— *Histoire de la philosophie, vol. I: L'antiquité et le Moyen Age. I: Introduction, Période hellénique* (Paris: Alcan, 1926).

—— *Histoire de la philosophie, vol. I: L'antiquité et le Moyen Age. II: Période hellénistique et romaine* (Paris: Alcan, 1927).

—— *Histoire de la philosophie, vol. I: L'antiquité et le Moyen Age. III: Moyen Age et Renaissance*, 1928).

—— *Histoire de la philosophie, vol. II: La philosophie moderne. I: Le dix-septième siècle* (Paris: Alcan, 1929).

—— *Histoire de la philosophie, vol. II: La philosophie moderne. II: Le dix-huitième siècle* (Paris: Alcan, 1930).

—— *Histoire de la philosophie, vol. II: La philosophie moderne. III: le XIXe siècle. Période des systèmes (1800–1950)* (Paris: Alcan, 1932).

—— *Histoire de la philosophie, vol. II: La philosophie moderne. IV: Le XIXe siècle après 1950. Le XXe siècle* (Paris: Alcan, 1932).

—— *La philosophie et son passé* (Paris: Presses universitaires de France, 1950 [1940]).

—— 'Les postulats de l'histoire de la philosophie', *Revue philosophique* 50 (1925): 48–78.

—— 'L'histoire de la philosophie dans l'enseignement', *Les Etudes philosophiques*, nouvelle série, 3/4 (1946): 211–17.

—— *Transformation de la philosophie française* (Paris: Flammarion, 1950).

—— 'Y a-t-il une philosophie chrétienne?', *Revue de métaphysique et de morale* 38 (1931): 133–62.

Bremond, André, 'Quelques essais de "religion" rationaliste: Alain, J. Bois, L. Brunschvicg, E. Gilson, D. Parodi, Th. Ruyssen', *Archives de philosophie* 8/4 (1931): 60–117.

—— 'Rationalisme et religion', *Archives de philosophie* 11/4 (1935): 1–203.

Brenner, Anastasios, *Les origines françaises de la philosophie des sciences* (Paris: Presses Universitaires de France, 2003).

—— 'Réconcilier les sciences et les lettres: Le rôle de l'histoire des sciences selon Paul Tannery, Gaston Milhaud et Abel Rey', *Revue d'histoire des sciences* 58 (2005): 433–54.

Brillouin, Léon, 'La crise de la physique moderne', *Revue philosophique* 49 (1924): 11–66.

Brooks, John I., *The Eclectic Legacy: Academic Philosophy and the Human Sciences in Nineteenth-Century France* (Newark and London: University of Delaware Press and Associated University Press, 1998).

Brosse, J., 'Gaston Bachelard. Portrait', *L'Arc* 20 (1962): 92–6.

Bruhl, Adrien, '[Hommage à Henri Berr]', *Revue de synthèse* 85 (1964): 54.

Brunschvicg, Léon, *De la connaissance de soi* (Paris: Alcan, 1931).
—— 'Du XIXe au XXe siècle, Troisième partie: L'humanisme de l'intelligence', *Revue de Paris* (February 1927): 576–91.
—— *Ecrits philosophiques* (Paris: Presses Universitaires de France, 1951).
—— '*Essai sur la connaissance approchée*, par Gaston Bachelard', *Revue philosophique* 54 (1929): 95–101.
—— '*Etude sur l'évolution d'un problème de physique. La propagation thermique dans les solides*, par Gaston Bachelard', *Revue philosophique* 54 (1929): 92–4.
—— 'History and Philosophy', in Raymond Klibansky (ed.), *Philosophy and History. Essays presented to Ernst Cassirer* (Oxford: Clarendon, 1936), pp. 27–34.
—— *Introduction à la vie de l'esprit* (Paris: Alcan, 1900).
—— 'La logique de Spinoza', *Revue de métaphysique et de morale* 1 (1893): 453–67.
—— *La modalité du jugement* (Paris: Alcan, 1897).
—— *La modalité du jugement, troisième édition augmentée de La vertu métaphysique du syllogisme selon Aristote, thèse latine traduite par Yvon Belaval* (Paris: Presses Universitaires de France, 1964).
—— 'La philosophie dans l'enseignement supérieure à Paris', *Annales de l'Université de Paris* 5 (1930): 215–23.
—— 'La révolution cartésienne et la notion spinoziste de la substance', *Revue de métaphysique et de morale* 2 (1904): 755–98.
—— *Les âges de l'intelligence* (Paris: Alcan, 1934).
—— *L'esprit européen* (n.p.: La Baconnière, 1947).
—— *Les étapes de la philosophie mathématique* (Paris: Alcan, 1912).
—— *L'expérience humaine et la causalité physique* (Paris: Alcan, 1922).
—— *L'idéalisme contemporain* (Paris: Alcan, 1921).
—— 'L'idée critique et le système kantien', *Revue de métaphysique et de morale* (1924): 133–203.
—— 'L'orientation du rationalisme: Représentation, concept, jugement', *Revue de métaphysique et de morale* 27 (1920): 261–343.
—— 'Mathématique et métaphysique chez Descartes', *Revue de métaphysique et de morale* 34 (1927): 277–324.
—— 'Nouvelles études sur l'anime primitive', *Revue des deux mondes* 52 (1932): 172–202.
—— 'Pascal savant', *Revue philosophique* 49 (1924): 5–27.
—— 'Religion et philosophie', *Revue de métaphysique et de morale* 42 (1935): 1–13.
—— *Spinoza* (Paris: Alcan, 1894).
—— 'Spinoza et ses contemporains', *Revue de métaphysique et de morale* 13 (1905): 673–705.
—— 'Spinoza et ses contemporains (suite)', *Revue de métaphysique et de morale* 14 (1906): 35–82.
—— 'Spinoza et ses contemporains (suite et fin)', *Revue de métaphysique et de morale* 14 (1906): 691–732.
—— *Spinoza et ses contemporains* (Paris: Alcan, 1923).
—— *Un Ministère de l'Education Nationale* (Paris: Plon, 1922).

—— 'Une phase du développement de la pensée mathématique', *Revue de métaphysique et de morale* 18 (1909): 309–56.

Brunschvicg, Léon and Pierre Boutroux (eds), *Œuvres de Blaise Pascal, publiées suivant l'ordre chronologique avec documents complémentaires, introductions et notes*, vols 1–3 (Paris: Hachette, 1908).

Brunschvicg, Léon, Pierre Boutroux and Félix Gazier (eds), *Œuvres de Blaise Pascal, publiées suivant l'ordre chronologique avec documents complémentaires, introductions et notes*, vols 4–6 (Paris: Hachette, 1914).

Brunschvicg, Léon et al., 'Histoire et Philosophie. Séance du 31 mai 1923', *Bulletin de la Société française de Philosophie* 23 (1923): 145–72.

—— 'La querelle de l'athéisme. Séance du 24 mars 1928', *Bulletin de la Société française de philosophie* 28/3 (1928): 50–95.

Bucciantini, Massimo, 'George Sarton e Aldo Mieli: bibliografia e concezioni della scienza a confronto', *Nuncius. Annali di storia della scienza* 2/2 (1987): 229–39.

Burguière, André, 'La notion de mentalité chez Marc Bloch and Lucien Febvre: deux conceptions, deux filiations', *Revue de synthèse* 111 (1983): 333–48.

Candar, Gilles and Jacqueline Pluet-Despatin, 'Introduction: une amitié au service de l'histoire', in Gilles Candar and Jacqueline Pluet-Despatin (eds), *Lucien Febvre: Lettres à Henri Berr* (Paris: Fayard, 1997), pp. i–xxxvi.

Canguilhem, Georges, *Ecrits sur la médicine* (Paris: Seuil, 2002).

—— *Etudes d'histoire et de philosophie des sciences concernant les vivants et la vie* (Paris: Vrin, 1994 [1968]).

—— *Idéologie et rationalité dans l'histoire des sciences de la vie* (Paris: Vrin, 1993 [1977]).

—— *La formation du concept de réflexe aux XVIIe et XVIIIe siècles* (Paris: Presses Universitaires de France, 1955).

—— *La vie et mort de Jean Cavaillès* (Paris: Allia, 1996 [1976]).

—— *Le normal et le pathologique* (Paris: Presses Universitaires de France, 1999 [1966]).

—— *The Normal and the Pathological* (trans. Carolyn R. Fawcett; New York: Zone Books, 1989).

—— 'Preface', *History and Technology* 4 (1987): 7–10.

Canguilhem, Georges and Camille Planet, *Traité de Logique et de Morale* (Marseille: Robert & Fils, 1939).

Caritat, Marie Jean Antoine Nicolas, Marquis de Condorcet, *Esquisse d'un tableau historique des progrès de l'esprit humain. Texte revu et présenté par O. H. Prior. Nouvelle édition présentée par Yvon Belaval* (Paris: Vrin, 1970 [1795]).

Castelão-Lawless, Teresa, 'Gaston Bachelard et le milieu scientifique et intellectuel français', in Pascal Nouvel (ed.), *Actualité et postérités de Gaston Bachelard* (Paris: Presses Universitaires de France, 1997), pp. 101–15.

—— 'Phenomenotechnique in Historical Perspective: Its Origins and Implications for Philosophy of Science', *Philosophy of Science* 62 (1995): 44–59.

Cazeneuve, Jean, *Lucien Lévy-Bruhl: sa vie, son œuvre* (Paris: Presses Universitaires de France, 1963).

Centre International de Synthèse, 'Le répertoire méthodique de synthèse scientifique', *Archeion* 9 (1928): 510–11.

—— *L'Hôtel de Nevers et le Centre International de Synthèse* (Paris: La Renaissance du livre, 1929).

Charle, Christophe, 'Bergson (Henri)', in Jacques Julliard and Michel Winock (eds), *Dictionnaire des intellectuels français: les personnes, les lieux, les moments* (Paris: Seuil, 1996), pp. 142–4.

—— *Dictionnaire biographique des universitaires aux XIXe et XXe siècles, vol. 1: La Faculté des Lettres de Paris (1809–1908)* (Paris: CNRS, 1985).

—— *La république des universitaires, 1870–1940* (Paris: Editions de Seuil, 1994).

—— 'Le beau marriage d'Emile Durkheim', *Actes de la recherche en sciences sociales* 55 (1984): 45–9.

—— 'Le Collège de France', in Pierre Nora (ed.), *Les Lieux de mémoire, II: La Nation* (Paris: Gallimard, 1986), vol. 3, pp. 389–424.

—— *Les professeurs de la Faculté des Lettres de Paris. Dictionnaire biographique 1909–1939* (Paris: Institut national de Recherche Pédagogique/Editions du CNRS, 1986).

Charle, Christophe and Eva Telkès, *Les professeurs du Collège de France. Dictionnaire biographique (1901–1939)* (Paris: CNRS, 1988).

Charrier, E., *Evolution intellectuelle féminine* (Paris: Mechelinck, 1931).

Chartier, Roger, *Cultural History: Between Practices and Representations* (trans. Lydia G. Cochrane; Ithaca: Cornell University Press, 1988).

Chatelet, François, *La philosophie des professeurs* (Paris: Grasset, 1970).

Chimisso, Cristina, 'From Phenomenology to Phenomenotechnique: The Role of Early-Twentieth-Century physics in Gaston Bachelard's Philosophy', *Studies in History and Philosophy of Science* (forthcoming).

—— *Gaston Bachelard: Critic of Science and the Imagination* (London: Routledge, 2001).

—— 'Hélène Metzger: The History of Science between the Study of Mentalities and Total History', *Studies in History and Philosophy of Science Part A*, 32/2 (2001): 203–41.

—— 'The Tribunal of Philosophy and its Norms: History and Philosophy in Georges Canguilhem's Historical Epistemology', *Studies in History and Philosophy of Biological and Biomedical Sciences* 34/2 (2003): 297–327.

Chimisso, Cristina and Gad Freudenthal, 'A Mind of Her Own: Hélène Metzger to Emile Meyerson, 1933', *Isis* 94/3 (2003): 477–91.

Choe, H., *Gaston Bachelard. Epistemologie. Bibliographie* (Frankfurt am Main: Lang, 1994).

Clark, Frances I., *The Position of Women in Contemporary France* (London: P.S. King & Son, 1937).

Clark, Terry N., *Prophets and Patrons: The French University System and the Emergence of the Social Sciences* (Cambridge, MA: Harvard University Press, 1973).

Cohen, Bernard I., 'Alexandre Koyré in America: Some Personal Reminiscences', *History and Technology* 4 (1987): 55–70.

Cohen, Bernard I. and René Taton, 'Hommage à Alexandre Koyré', in Bernard I. Cohen and René Taton (eds), *Mélanges Alexandre Koyré, I: L'aventure de la science* (Paris: Hermann, 1964), pp. xix–xxv.

Collège de France website, 'Liste des professeurs depuis la fondation du Collège de France en 1530' http://www.college-de-france.fr/media/lis_prf/UPL45507_LISTE_DES_PROFESSEURS.pdf (accessed 14 November 2007).

Comité International d'Histoire des Sciences, 'Séance de lundí, 27 septembre 1937', *Archeion* 19 (1937): 382–8.

—— 'Séance privée du Comité (14 mai, après midi) à la Sorbonne', *Archeion* 14 (1932): 468–73.

Comité international et Centre international d'histoire des sciences, 'Election de nouveaux membres', *Archeion*, 12 (1930): 45.

—— 'Listes des membres du Comité International d'Histoire des Sciences', *Archeion* 11 (1929): 84–6.

Contributori di Wikipedia, 'Aldo Mieli', *Wikipedia, L'enciclopedia libera* http://it.wikipedia.org/wiki/Aldo_Mieli (accessed 30 April 2007).

Corsini, Andrea, 'Aldo Mieli', *Rivista di storia delle scienze mediche e naturali* 41/1 (1950): 111–13.

Couturat, Louis, 'IIme Congrès de Philosophie, Genève, comptes rendus critiques: II Logique et philosophie des sciences', *Revue de métaphysique et de morale* 12 (1904): 1037–1077.

Cresson, André, *Les courants de la pensée philosophique française* (Paris: Colin, 1927).

—— 'Lettre sur Léon Brunschvicg', *Revue de métaphysique et de morale* 50 (1945): 5–7.

Dagognet, François, *Gaston Bachelard: sa vie, son œuvre* (Paris: Presses Universitaires de France, 1965).

—— *Georges Canguilhem: Philosophe de la vie* (Le Plessis-Robinson: Institut Synthélabo, 1997).

—— 'M. Brunschvicg et Bachelard', *Revue de métaphysique et de morale* 70 (1965): 43–54.

—— 'Une œuvre en trois temps', *Revue de métaphysique et de morale* 90 (1985): 29–38.

Davis, Natalie Zemon, 'Women and the World of the *Annales*', *History Workshop Journal* 33 (1992): 121–37.

Davy, Georges, 'Gaston Bachelard: l'unité de l'homme et de l'œuvre', *Etudes philosophiques* 2 (1952): 123–33.

—— 'Le problème de l'industrialisation de l'Etat', *Revue de métaphysique et de morale* 31 (1924): 599–641.

—— *Sociologues d'hier et d'aujourd'hui* (Paris: Alcan, 1931).

De Broglie, Louis, 'Le déterminisme et la causalité dans la physique contemporaine. Séance du 12 novembre 1929', *Bulletin de la Société française de Philosophie* 29/5 (1929): 141–60.

Debru, Claude, *Georges Canguilhem, science et non-science* (Paris: Editions rue d'Ulm/Presses de l'Ecole normale supérieure, 2004).

—— 'L'engagement philosophique dans le champ de la médicine: Georges Canguilhem aujourd'hui', in Jean-François Braunstein (ed.), *Canguilhem: Histoire des sciences et politique du vivant* (Paris: Presses Universitaires de France, 2007), pp. 45–62.

Delacroix, H., 'Le fonctionnement psychologique du language', *Revue de métaphysique et de morale* 31 (1924): 1–17.

—— 'Les conditions psychologiques du langage', *Revue philosophique* 49 (1924): 28–66.

Delaporte, François (ed.), *A Vital Rationalist: Selected Writings from Georges Canguilhem* (New York: Zone Books, 1994).

Delbos, Victor, 'De la méthode en histoire de la philosophie, I: Les matériaux de la reconstitution historique des doctrines', *Revue de métaphysique et de morale* 24 (1917): 279–89.

—— 'De la méthode en histoire de la philosophie, II: Analyse et reconstitution des doctrines', *Revue de métaphysique et de morale* 24 (1917): 369–82.

—— *Des Kant aux postkantiens* (Paris: Aubier, 1992 [1940]).

—— *La philosophie française* (Paris: Plon, 1919).

—— 'Les conceptions de l'histoire de la philosophie', *Revue de métaphysique et de morale* 24 (1917): 135–47.

—— 'Les facteurs kantiens de la philosophie allemande du commencement du XIXe siècle', *Revue de métaphysique et de morale* 26 (1919): 569–93.

—— 'Les facteurs kantiens de la philosophie allemande de la fin du XVIIIe siècle et du commencement du XIXe', *Revue de métaphysique et de morale* 27 (1920): 1–25.

—— 'Les facteurs kantiens de la philosophie allemande de la fin du XVIIIe siècle et du commencement du XIXe', *Revue de métaphysique et de morale* 28 (1921): 27–47.

—— 'Les facteurs kantiens de la philosophie allemande de la fin du XVIIIe siècle et du commencement du XIXe', *Revue de métaphysique et de morale* 29 (1922): 157–76.

Deleuze, Gilles, *Bergsonism* (trans. Hugh Tomlison and Barbara Habberjam; New York: Zone Books, 1991 [1966]).

Delorme, Suzanne, 'Hommage à Alexandre Koyré', *Revue d'histoire des sciences* 18/2 (1965): 129–39.

Deschoux, Marcel, *Léon Brunschvicg ou l'idéalisme à hauteur d'homme* (Paris: Seghers, 1969).

Devolvé, J. et al., 'L'efficacité des doctrines morales. Séance du 20 mai 1909', *Bulletin de la Société française de Philosophie* 9 (1909): 193–233.

Di Meo, Antonio, 'Aldo Mieli e la storia della chimica in Italia', *Archives internationales d'histoire des sciences* 36/117 (1986): 337–61.

Di Palma, Patrizia, *Lucien Lévy-Bruhl: dalla scienza dei costumi all'antropologia* (Lucca: Pacini, 1983).

Doléris, Dr [Jacques-Amédée] et al., 'L'éducation sexuelle. Séance du 18 février 1911', *Bulletin de la Société française de Philosophie* 11 (1911): 29–51.

Dry, Sarah, *Curie* (London: Haus, 2003).

Duhem, Pierre, *La théorie physique. Son objet, sa structure. Deuxième édition revue et augmentée* (Paris: Vrin, 1981 [1914]).

—— *Le Système du Monde: Histoire des doctrines cosmologiques de Platon à Copernic*. (10 vols; Paris: Hermann, 1913–59).

Dumas, Georges, 'Lévy-Bruhl, La philosophie d'Auguste Comte', *Revue philosophique* 25 (1900): 396–407.

—— (ed.), *Traité de psychologie* (Paris: Alcan, 1923–24).

Durkheim, Emile, *The Elementary Forms of Religious Life* (trans. Karen E. Fields; New York: The Free Press, 1995 [1912]).

—— 'L'enseignement philosophique et l'agrégation de philosophie', *Revue philosophique* 39 (1895): 121–47.

—— *Les formes élémentaires de la vie religieuse. Le système totémique en Australie* (Paris: Alcan, 1912).

—— 'Les jugements de valeur et les jugements de réalité', *Revue de métaphysique et de morale* 19 (1911): 437–53.

—— *L'évolution pédagogique en France* (Paris: Presses Universitaires de France, 1990 [1938]).

—— 'Lévy-Bruhl, La morale et la sciences des mœurs', *Année sociologique* 7 (1902–1903): 380–84.

—— 'Pédagogie et sociologie', *Revue de métaphysique et de morale* 11 (1903): 37–54.

—— 'Représentations individuelles et représentations collectives', *Revue de métaphysique et de morale* 6 (1898): 273–302.

—— 'Sociologie religieuse et théorie de la connaissance', *Revue de métaphysique et de morale* 17 (1909): 733–58.

Durkheim, Emile and Marcel Mauss, 'De quelques formes primitive de classification: contribution à l'étude des représentations collectives', *L'année sociologique* 6 (1901–1902): 1–72.

—— *Primitive Classification* (trans. Rodney Needham; London: Cohen & West, 1969 [1901–1902]).

Durkheim, Emile et al., 'La détermination du fait moral. Séance du 11 février 1906', *Bulletin de la Société française de Philosophie* 6 (1906): 113–67.

—— 'La détermination du fait moral (suite). Séance du 22 mars 1906', *Bulletin de la Société française de Philosophie* 6 (1906): 169–211.

—— 'Le problème religieux et la dualité de la nature humaine. Séance du 4 février 1913', *Bulletin de la Société française de Philosophie* 13 (1913): 63–113.

Duvignaud, Jean, *Le langage perdu. Essai sur la différence anthropologique* (Paris: Presses Universitaires de France, 1973).

Ecole normale supérieure, 'Rapport du directeur', *Annales de l'Université de Paris* 2 (1927): 213–19.

Ecole normale supérieure website, 'Présentation de l'ENS' http://www.ens.fr/ecole/presentation.php (accessed 2 April 2007).

Einstein, Albert et al., 'La théorie de la relativité. Séance du 6 avril 1922', *Bulletin de la Société française de Philosophie* 22/3 (1922): 91–113.

Elkana, Yehuda, 'Alexandre Koyré: Between the History of Ideas and Sociology of Disembodied Knowledge', *History and Technology* 4/1–4 (1987): 115–48.

Enriques, Federigo, 'La signification et l'importance de l'histoire de la science et l'œuvre de Paul Tannery', *Revue de métaphysique et de morale* 31 (1924): 425–34.

Fabiani, Jean-Louis, 'Brunschvicg (Léon)', in Jacques Julliard and Michel Winock (eds), *Dictionnaire des intellectuels français: les personnes, le lieux, les moments* (Paris: Seuil, 1996), pp. 197–8.

—— *Les philosophes de la République* (Paris: Editions de Minuit, 1988).

—— 'Les programmes et les œuvres: professeurs de philosophie en classe et en ville au tournant du siècle', *Actes de la recherche en sciences sociales* 47–48 (1983): 3–20.

Faculté des Lettres, 'Rapport annuel du Doyen: année scolaire 1925–26', *Annales de l'Université de Paris* 2/4 (1927): 295–309.

—— 'Rapport annuel du Doyen: année scolaire 1927–28', *Annales de l'Université de Paris* 4/6 (1929): 445–73.

—— 'Rapport annuel du Doyen: année scolaire 1933–34', *Annales de l'Université de Paris* 10/5 (1935): 385–404.

Febvre, Lucien, *Combats pour l'histoire* (Paris: Colin, 1992).

—— *De la Revue de synthèse aux Annales. Lettres à Henri Berr 1911–1954. Etablissement du texte, présentation et notes par Gilles Candar et Jacqueline Pluet-Despatin* (Paris: Fayard, 1997).

—— 'De l'à peu près à la précision en passant par l'ouï-dire', *Annales E.S.C.* 5 (1950): 25–31.

—— *Le problème de l'incroyance au XVIe siècle: la religion de Rabelais* (Paris: Albin Michel, 1942).

—— *The Problem of the Unbelief in the Sixteenth Century: The Religion of Rabelais* (trans. Beatrice Gottlieb; Cambridge, MA: Harvard University Press, 1982 [1942]).

—— 'Une encyclopédie française: pourquoi, comment?', in Lucien Febvre (ed.), *Encyclopédie française* (Paris: Société de Gestion de l'Encyclopédie Française, 1937), 1.04.11–1.04.14.

Febvre, Lucien and Lionel Bataillon, *La terre et l'évolution de l'humanité: Introduction géographique à l'histoire* (Paris: La Renaissance du livre, 1922).

Febvre, Lucien and Henri-Jean Martin, *L'apparition du livre* (Paris: Albin Michel, 1958).

Fichant, Michel, 'Georges Canguilhem et l'idée de la philosophie', in Etienne Balibar et al. (eds), *Georges Canguilhem: Philosophe, historien des sciences. Actes du colloque (6–7–8 décembre 1990)* (Paris: Albin Michel, 1993), pp. 37–48.

—— 'L'épistémologie en France', in François Chatelet (ed.), *Histoire de la philosophie: le 20e siècle* (Paris: Hachette, 1973), pp. 135–78.

Fimiani, Mariapaola, *L'arcaico e l'attuale: Lévy-Bruhl, Mauss, Foucault* (Torino: Bollati Boringhieri, 2000).

Finocchiaro, M., 'Logic and Scholarship in Koyré's historiography', *Physis* 19 (1977): 5–27.

Fondation 'Pour la science', *Bulletin du Centre International de Synthèse. Section de synthèse historique* (Paris: Appendix of *Revue de synthèse* 42, nouvelle série 16, 1926).

Foucault, Michel, *The Archaeology of Knowledge* (trans. A. M. Sheridan Smith; London: Tavistock, 1972 [1969]).

—— *The Birth of the Clinic: An Archaeology of Medical Perception* (trans. A.M. Sheridan Smith; London: Routledge, 2003 [1963]).

—— 'La vie: l'expérience et la science', *Revue de métaphysique et de morale* 90 (1985): 3–14.

Fournier, Marcel, *Marcel Mauss* (trans. Jane Marie Todd; Princeton and Oxford: Princeton University Press, 2006 [1994]).

Frazer, Sir James George, *The Golden Bough: A Study in Comparative Religion* (2 vols; London: Macmillan & Co., 1890).

—— *Myths of the Origin of Fire* (London: Macmillan & Co., 1930).

Freud, Sigmund, *Civilization and its Discontents* (trans. David McLintock; London: Penguin, 2002 [1930]).

Freudenthal, Gad, 'Epistémologie des sciences de la nature et herméneutique de l'histoire des sciences selon Hélène Metzger', in Gad Freudenthal (ed.), *Etudes sur/Studies on Hélène Metzger* (Leiden: Brill, 1990), pp. 161–88.

—— 'Hélène Metzger (1888–1944)', in Michel Bitbol and Jean Gayon (eds), *L'épistémologie française, 1830–1970* (Paris: Presses Universitaires de France, 2006), pp. 113–56.

—— 'Hélène Metzger: Eléments de biographie', in Gad Freudenthal (ed.), *Etudes sur/Studies on Hélène Metzger* (Leiden: Brill, 1990), pp. 197–208.

Fruteau de Laclos, Frédéric, 'La philosophie de l'intellect d'Emile Meyerson. De l'épistémologie à la psychologie' (unpublished thesis: Université de Paris-X Nanterre, 2004).

Fugler, Martin, 'Fondateurs et collaborateurs, les débuts de la *Revue de synthèse historique*', in Agnès Biard, Dominique Bourel and Eric Brian (eds), *Henri Berr et la culture du XXe siècle. Histoire, science et philosophie* (Paris: Albin Michel/ Centre international de synthèse, 1997), pp. 173–88.

Gagey, Jacques, *Gaston Bachelard ou la conversion a l'imaginaire* (Paris: Rivière, 1969).

Gayon, Jean, 'The Concept of Individuality in Canguilhem's Philosophy of Biology', *Journal of the History of Biology* 31 (1998): 205–325.

Gayon, Jean and Jean-Jacques Wunenburger (eds), *Bachelard dans le monde* (Paris: Presses Universitaires de France, 2000).

Geldsetzer, Lutz, *Bibliography of the International Congresses of Philosophy/ Bibliographie der Internationales philosophie Kongresse: Proceedings/Beiträge 1900–1978* (Munich, New York, London and Paris: Saur, 1981).

Gemelli, Giuliana, 'Communauté intellectuelle et stratégies institutionnelles. Henri Berr et la fondation du Centre International de Synthèse', *Revue de synthèse*, IV series, 2 (1987): 225–59.

Gilson, Etienne, 'Concerning Christian Philosophy: The Distinctiveness of the Philosophic Order', in Raymond Klibansky (ed.), *Philosophy and History* (Oxford: Clarendon Press, 1936), pp. 61–76.

Gilson, Etienne et al., 'La notion de philosophie chrétienne. Séance du 21 mars 1931', *Bulletin de la Société française de Philosophie* 31 (1931): 37–93.

Ginestier, Paul, *Pour connaître Bachelard* (Paris: Bordas, 1987).

Gliozzi, Mario, 'Appendice ad Praefatione de volume XXL', *Archeion* 21 (1938): 10–12.

Golinski, Jan V., 'Hélène Metzger and the Interpretation of Seventeenth Century Chemistry', *History of Science* 25/1 (1987): 85–97.

Goyard-Fabre, Simone, 'Léon Brunschvicg et l'héritage kantien', in Jean Ferrari, Margit Ruffing, Robert Theis and Matthias Vollet (eds), *Kant et la France – Kant und Frankreich* (Hildesheim, Zurich and New York: Georg Olms, 2005), pp. 261–9.

Gracia, Jorge J.E., 'The Enlightening Gloss: Gilson and the History of Philosophy', in Peter A. Redpath (ed.), *A Thomistic Tapestry: Essay in Memory of Etienne Gilson* (Amsterdam and New York: Rodopi B.V., 2003), pp. 1–11.

Greco, Monica, 'Between Social and Organic Norms: Reading Canguilhem and "Somatization" ', *Economy and Society* 27/2–3 (1998): 234–48.

Gueroult, Martial, *Dianoématique, livre II: Philosophie de l'histoire de la philosophie* (Paris: Aubier, 1979).

—— *Dianoématique, livre I: L'histoire de l'histoire de la philosophie, 3: en France de Condorcet à nos jours* (Paris: Aubier, 1988).

—— 'The History of Philosophy as a Philosophical Problem', *The Monist* 53 (1969): 563–87.

Gueroult, Martial et al., 'Commémoration du Xe anniversaire de la mort de Léon Brunschvicg: Brunschvicg et l'Histoire de la Civilisation. Séance du 30 janvier 1954', *Bulletin de la Société française de Philosophie* 48 (1954): 1–36.

Guigue, Albert, *La Faculté des Lettres de l'Université de Paris depuis sa fondation (17 mars 1808) jusqu'au 1er Janvier 1935* (Paris: Alcan, 1935).

Gutting, Gary, *French Philosophy in the Twentieth Century* (Cambridge: Cambridge University Press, 2001).

—— 'French Philosophy of Science', in by E. Craig (ed.), *Routledge Encyclopedia of Philosophy* (London: Routledge, 1998) Available on-line at http://www.rep.routledge.com/article/Q038 (accessed 9 October 2006).

—— 'Introduction: What is Continental Philosophy of Science?', in Gary Gutting (ed.), *Continental Philosophy of Science* (Oxford: Blackwell, 2005), pp. 1–16.

—— *Michel Foucault's Archaeology of Scientific Reason* (Cambridge: Cambridge University Press, 1989).

Hacking, Ian, *Representing and Intervening* (Cambridge: Cambridge University Press, 1983).

Heidelberger, Michael, 'History of Science and Criticism of Positivism: Emile Meyerson's and Hélène Metzger's Views from a Present-day Perspective', in Gad Freudenthal (ed.), *Etudes sur/Studies on Hélène Metzger* (Leiden: Brill, 1990), pp. 151–60.

Horton, Robin, 'Lévy-Bruhl, Durkheim and the Scientific Revolution', in Robin Horton and Ruth Finnegan (eds), *Modes of Thought. Essays on Thinking in Western and Non-Western Societies* (London: Faber & Faber1973), pp. 249–305.

Horton, Robin and Ruth Finnegan (eds), *Modes of Thought. Essays on Thinking in Western and non-Western societies* (London: Faber & Faber, 1973).

Humbert, Pierre, *Pierre Duhem* (Paris: Bluod et Gay, 1932).

Hyppolite, Jean, *Introduction à la philosophie de l'histoire de Hegel* (Paris: Rivière, 1948).

—— 'Les travaux de jeunesse de Hegel d'après des ouvrages récents', *Revue de métaphysique et de morale* 42 (1935): 399–426.

—— 'Les travaux de jeunesse de Hegel d'après des ouvrages récents (suite et fin)', *Revue de métaphysique et de morale* 42 (1935): 549–78.

Institut de France website, http://www.institut-de-france.fr/institut/index.html (accessed 20 July 2006).

International Congress of Philosophy (ed.), *Atti del IV Congresso Internazionale di filosofia, Bologna MCMXI* (3 vols; Genova: Formiggini, 1912–16).

International Federation of University Women, *Lexique international des termes universitaires* (Paris, London: Fédération Internationale des femmes diplômées des universités, 1939).

Jaki, S.T., *Uneasy Genius: The Life and the Work of Pierre Duhem* (Dordrecht: Martinus Nijhoff, 1987).

Jambet, Christian (ed.), *Henry Corbin* (Paris: Editions de l'Herne, 1981).

Janet, Paul, *Victor Cousin et son œuvre* (Paris: Calmann Lévy, 1885).

Jardine, Nicholas, 'Koyré's Intellectual Revolution', *La lettre de la Maison française d'Oxford* 13 (2001): 11–25.

—— 'Koyré's Kepler/Kepler's Koyré', *History of Science* 38 (2000): 363–77.

Jorland, Gérard, 'La notion de révolution scientifique: la modèle de Kuhn', in Michel Bitbol and Jean Gayon (eds), *L'épistémologie française, 1830–1970* (Paris: Presses Universitaires de France, 2006), pp. 157–71.

—— *La science dans la philosophie: les recherches épistémologiques d'Alexandre Koyré* (Paris: Gallimard, 1981).

Julliard, Jacques and Michel Winock (eds), *Dictionnaire des intellectuels français: les personnes, les lieux, les moments* (Paris: Seuil, 1996).

Karady, Victor, 'The Durkheimian Sociology in Academe. A Reconsideration', in Philippe Besnard (ed.), *The Sociological Domain: The Durkheimians and the Founding of French Sociology* (Cambridge: Cambridge University Press, 1983), pp. 71–89.

Keck, Frédérick, 'Causalité mentale et perception de l'invisible. Le concept de participation chez Lévy-Bruhl', *Revue philosophique* 130 (2005): 304–322.

Kelly, Michael, *Hegel in France* (Birmingham: Birmingham Modern Languages Publications, 1992).

Kleinberg, Ethan, *Generation Existential: Heidegger's Philosophy in France 1927–1961* (Ithaca and London: Cornell University Press, 2005).

Köhnke, Klaus Christian, *The Rise of Neo-Kantianism* (Cambridge: Cambridge University Press, 1991).

Koyré, Alexandre, 'Au l'aurore de la science moderne. La jeunesse de Galilée (I)', *Annales de l'Université de Paris* 10 (1935): 540–51.

—— 'Commentary', in A.C. Crombie (ed.), *Scientific Change: Historical Studies in the Intellectual, Social and Technical Conditions for Scientific Discoveries and Technical Invention, from Antiquity to the Present* (London: Heinemann, 1963), pp. 847–57.

—— 'Copernic', *Revue philosophique* 58 (1933): 101–18.

—— *De la mystique à la science. Cours, conférences et documents 1922–1962, édité par Pietro Redondi* (Paris: Ed. de l'Ecole des Hautes Etudes, 1986).

—— *Essai sur l'idée de Dieu et les preuves de son existence chez Descartes* (Paris: Leroux, 1922).

—— *Etudes d'histoire de la pensée philosophique* (Paris: Gallimard, 1971).

—— *Etudes d'histoire de la pensée scientifique* (Paris: Presse Universitaires de France, 1966).

—— *Etudes galiléennes* (Paris: Hermann, 1939).

—— *From the Closed World to the Infinite Universe* (New York: Harper, 1958 [1957]).

—— 'Galilée et l'expérience de Pise: à propos d'une légende', *Annales de l'Université de Paris* 12 (1937): 442–53.

—— *Galileo Studies* (trans. John Mepham; Hassocks: The Harvest Press, 1978).

—— 'Hegel à Jena', *Revue philosophique* 59 (1934): 274–83.

—— 'Hegel en Russie', *Monde Slave* 1 (1936): 379–88.

—— 'Introduction [of M. Heidegger "Qu'est-ce que la métaphysique"]', *Bifor* 8 (1931): 5–8.

—— *La philosophie de Jacob Boehme. Etudes sur les origines de la métaphysique allemande* (Paris: Vrin, 1929).

—— *La philosophie et le problème national en Russie au début di XIXe siècle* (Paris: Honoré Champion, 1929).

—— '[Letter to Henry Corbin, 1937 or 1938]', in Christian Jambet (ed.), *Henry Corbin* (Paris: L'Herne, 1981), pp. 330.

—— *L'idée de Dieu dans la philosophie de saint Anselme* (Paris: Leroux, 1923).

—— *Metaphysics and Measurement: Essays in Scientific Revolution* (Cambridge, MA: Harvard University Press, 1968).

—— 'Note sur la langue et la terminologie hégéliennes', *Revue philosophique* 56 (1931): 409–39.

—— 'Rapport sur l'état des études hégéliennes en France', *Revue d'histoire de la philosophie* 5 (1931): 147–71.

Kragh, Helge, *Quantum Generations: A History of Physics in the Twentieth Century* (Princeton: Princeton University Press, 1999).

Kuhn, Thomas S., 'Alexandre Koyré and the History of Science', *Encounter* 34 (1970): 67–9.

—— *The Structure of Scientific Revolutions* (Chicago: The University of Chicago Press, 1996 [1962]).

Kusch, Martin, *Psychological Knowledge: A Social History and Philosophy* (London and New York: Routledge, 1999).

Lagache, Daniel, 'Le normal et le pathologique d'après M. Georges Canguilhem', *Revue de métaphysique et de morale* 51 (1946): 355–70.

Lalande, André, 'Dominique Parodi', *Revue de métaphysique et de morale* 60 (1955): 341–51.

—— 'La Philosophie en France: 1945–1946', *The Philosophical Review* 56/1 (1947): 1–18.

—— 'Philosophy in France, 1918', *The Philosophical Review* 28/5 (1919): 443–65.

—— *Vocabulaire technique et critique de la philosophie* (2 vols; Paris: Presses Universitaires de France, 1999 [1926]).

LaLumia, Joseph, *The Ways of Reason: A Critical Study of the Ideas of Emile Meyerson* (London: Allen & Unwin, 1966).
Larvor, Brendan, 'Why did Kuhn's *Structure of Scientific Revolutions* Cause a Fuss?', *Studies in History and Philosophy of Science* 34 (2003): 369–90.
Lautman, Jacques, 'Un stoïcien chaleureux', *Revue d'histoire des sciences* 53/1 (2000): 27–45.
Le Blanc, Guillaime, *Canguilhem et le normes* (Paris: Presses Universitaires de France, 1998).
—— *La vie humaine: anthropologie et biologie chez Georges Canguilhem* (Paris: Presses Universitaires de France, 2002).
Le Goff, Jacques, 'Histoire des sciences et histoire des mentalités', *Revue de synthèse* 104 (1983): 405–15.
Lecourt, Dominique, *Bachelard ou le jour et la nuit* (Paris: Grasset, 1974).
—— 'Georges Canguilhem, le philosophe', in Jean-François Braunstein (ed.), *Canguilhem: Histoire des sciences et politique du vivant* (Paris: Presses Universitaires de France, 2007), pp. 27–43.
—— *L'épistémologie historique de Gaston Bachelard* (Paris: Vrin, 1969).
—— *La philosophie des sciences* (Paris: Presses Universitaires de France, 2001).
—— *Marxism and Epistemology. Bachelard, Canguilhem and Foucault* (trans. Ben Brewster; London: NLB, 1975).
—— *Pour une critique de l'épistémologie* (Paris: Maspero, 1972).
Lefèvre, Frédéric, *Une heure avec ...* (Paris: Gallimard, 1924–27).
Leibniz, G.W., 'Trois lettres', *Revue philosophique* 49 (1924): 5–10.
Lévy, Isidore, *Recherches sur les sources de la légende de Pythagore* (Paris: Leroux, 1926).
Lévy-Bruhl, Lucien, *Carnets* (Paris: Presses Universitaires de France, 1949).
—— *Ethics and Moral Science* (trans. Elisabeth Lee; London: Archibald Constable, 1905 [1903]).
—— *History of Modern Philosophy in France* (London: Kegan Paul, Trench, Trübner & Co., 1899).
—— *Jean Jaurès. Essai biographique. Nouvelle édition suivie de lettres inédites* (Paris: Rieder, 1924).
—— *L'Allemagne depuis Leibniz. Essai sur le développement de la conscience nationale en Allemagne, 1700–1848* (Paris: Hachette, 1890).
—— *La mentalité primitive* (Paris: Alcan, 1922).
—— 'La mentalité primitive et les médicins européens', *Revue de Paris* 16 (1921): 806–26.
—— *La mentalité primitive: The Herbert Spencer Lecture, delivered at Oxford 29 May 1931* (Oxford: Clarendon Press, 1931).
—— *L'âme primitive* (Paris: Presses Universitaires de France, 1996 [1927]).
—— *La morale et la science des mœurs* (Paris: Alcan, 1903).
—— *La philosophie d'Auguste Comte* (Paris: Alcan, 1900).
—— *La philosophie de Jacobi* (Paris: Alcan, 1894).
—— 'Le centenaire d'Auguste Comte', *Revue des deux mondes* 145 (1898): 394–424.

—— *Le surnaturel et la nature dans la mentalité primitive* (Paris: Presses Universitaires de France, 1963 [1931]).

—— *Les fonctions mentales dans les sociétés inférieures* (Paris: Alcan, 1910).

—— *L'expérience mystique et les symboles chez les primitifs* (Paris: Alcan, 1938).

—— 'L'idéal républicain', *Revue de Paris* 31 (1924): 805–22.

—— *L'idée de responsabilité* (Paris: Hachette, 1884).

—— *Quid de Deo Seneca senserit* (Paris: Hachette, 1884).

—— (ed.), *Correspondance de John Stuart Mill et d'Auguste Comte* (Paris: Alcan, 1899).

Lévy-Bruhl, Lucien et al., 'La mentalité primitive. Séance du 15 février 1923', *Bulletin de la Société française de Philosophie* 23 (1923): 17–48.

—— 'L'âme primitive. Séance du 1er juin 1929', *Bulletin de la Société française de Philosophie* 29 (1929): 105–37.

Lloyd, G.E.R., *Demystifying Mentalities* (Cambridge: Cambridge University Press, 1990).

Löwy, Ilana, 'Constructivist epistemologies: Metzger and Fleck', in Gad Freudenthal (ed.), *Etudes sur/Studies on Hélène Metzger* (Leiden: Brill, 1990), pp. 219–35.

Macey, David, 'The Honour of Georges Canguilhem', *Economy and Society* 27/2–3 (1998): 171–81.

Margadant, Jo Burr, *Madame le professeur: Women Educators in the Third Republic* (Princeton, NJ: Princeton University Press, 1990).

Mauss, Marcel, 'Appréciation sociologique du bolchevisme', *Revue de métaphysique et de morale* 31 (1924): 103–32.

—— 'Lévy-Bruhl sociologue', *Revue philosophique* 64 (1939): 251–3.

McAllester Jones, Mary, *Gaston Bachelard, Subversive Humanist: Texts and Readings* (Madison, WI: University of Wisconsin Press, 1991).

McArthur, Dan, 'Why Bachelard is not a Scientific Realist', *Philosophical Forum* 33/2 (2002): 159–72.

Merllié, Dominique, 'Les rapports entre la *Revue de métaphysique* et la *Revue philosophique*: Xavier Léon, Théodule Ribot, Lucien Lévy-Bruhl', *Revue de métaphysique et de morale* 98 (1993): 59–108.

—— 'Lévy-Bruhl et Durkheim. Notes biographiques en marge d'une correspondance', *Revue philosophique* 114 (1989): 493–514.

—— 'Présentation. Le cas Lévy-Bruhl', *Revue philosophique* 114 (1989): 419–48.

Metzger, Hélène, 'Activité de la Section d'histoire des sciences en 1930–31', *Revue de synthèse* 51 (1931): 192–3.

—— *Attraction universelle et religion naturelle chez quelques commentateurs anglais de Newton. Première partie. Introduction philosophique* (Paris: Hermann, 1938).

—— *Attraction universelle et religion naturelle chez quelques commentateurs anglais de Newton. Deuxième partie. Newton–Bentley–Whiston–Toland* (Paris: Hermann, 1938).

—— *Attraction universelle et religion naturelle chez quelques commentateurs anglais de Newton. Troisième partie. Clarke–Cheyne–Derham–Baxter–Priestly* (Paris: Hermann, 1938).

—— 'Gaston Bachelard, *Essai sur la connaissance approchée*, Paris 1928; Gaston Bachelard, *La valeur inductive de la relativité*, Paris 1929; Gaston Bachelard, *Etude sur l'évolution d'un problème de physique*, Paris 1928', *Archeion* 12 (1930): 218–20.

—— 'Gaston Bachelard. *La Formation de l'esprit scientifique. Contribution à une psychanalyse de la connaissance objective*', *Archeion* 21 (1938): 162–5.

—— *La Chimie* (Paris: Boccard, 1930).

—— *La genèse de la science des cristaux* (Paris: Alcan, 1918).

—— 'La littérature chimique française aux 17e et 18e siècles', *Thalès* 2 (1935): 162–5.

—— 'La méthode philosophique dans l'histoire des sciences', *Archeion* 19 (1937): 204–16.

—— *La méthode philosophique en histoire des sciences. Textes 1914–1939, réunis par Gad Freudenthal* (Paris: Fayard, 1987).

—— *La philosophie de la matière chez Lavoisier* (Paris: Hermann, 1935).

—— 'La philosophie de la matière chez Stahl et ses disciples', *Isis* 8 (1926): 427–64.

—— 'La philosophie de Lévy-Bruhl et l'histoire des sciences', *Archeion* 12 (1930): 15–24.

—— 'La philosophie d'Emile Meyerson et l'histoire des sciences', *Archeion* 11 (1929): xxxii–xliii.

—— 'L'*a priori* dans la doctrine scientifique et l'histoire des sciences', *Archeion* 18 (1936): 29–42.

—— 'Le rôle des précurseurs dans l'évolution de la science', *Thalès* 4 (1939): 199–209.

—— 'Léon Brunschvicg, *De la connaissance de soi*', *Archeion* 13 (1931): 389–91.

—— 'Léon Brunschvicg, *La physique du vingtième siècle et la philosophie; Le rôle du pythagorisme dans l'évolution des idées*', *Archeion* 19 (1937): 301–303.

—— 'Léon Brunschvicg, *Les âges de l'intelligence*', *Archeion* 16 (1934): 254–5.

—— 'Léon Brunschvicg, *L'expérience humaine et la causalité physique*', *Isis* 5 (1923): 479–83.

—— *Les concepts scientifiques* (Paris: Alcan, 1926).

—— *Les doctrines chimiques en France du début du XVIIe à la fin du XVIIIe siècle* (Paris: Presses Universitaires de France, 1969 [1923]).

—— 'L'historien des sciences doit-il se faire le contemporain des savants dont il parle?' *Archeion* 15 (1933): 34–44.

—— 'Lucien Lévy-Bruhl, *L'âme primitive*', *Isis* 9 (1927): 482–6.

—— *Newton, Stahl, Boerhaave et la doctrine chimique* (Paris: Alcan, 1974 [1930]).

Metzger, Paul, *Le conseil supérieur et le grand bailliage de Lyon* (Lyon: Université de Lyon, 1913).

Meyerson, Emile, *Du cheminement de la pensée* (Paris: Alcan, 1931).

—— 'La tendance apriorique et l'expérience', *Revue philosophique*, 49 (1924): 161–79.

—— 'Le relativisme, théorie du réel', *Revue de métaphysique et de morale* 31 (1924): 29–48.

Mieli, Aldo, 'Bibliografia metodica dei lavori di storia della scienza pubblicati in Italia', *Archivio di storia della scienza* 1 (1919): 84–6.

—— 'Digressions autobiographiques sous forme de préface à un panorama général d'Histoire des Sciences', *Archeion* 37 (1947–48): 494–505.

—— 'Hélène Metzger, *La genèse de la sciences des cristaux*; *Les doctrines chimiques en France du début du XVII à la fin du XVII siècle*', *Archivio di storia della scienza* 4 (1923): 290–91.

—— 'Hélène Metzger, *Les concepts scientifiques*', *Archeion*, 8 (1927): 270–71.

—— 'Herbertz Richard, *Philosophie und Einzelwissenschaften*', *Rivista di filosofia* 5 (1913): 300–302.

—— 'Il tricentenario dei "Discorsi" di Galileo Galilei', *Archeion* 21 (1938): 193–297.

—— *La science arabe et son rôle dans l'évolution scientifique mondiale. Avec quelques additions de Henri-Paul-Joseph Renaud, Max Meyerhof, Julius Ruska* (Leiden: Brill, 1938).

—— 'La storia della scienza in Italia (Prolusione ad un corso di Storia delle science, tenuta nella R. Università di Perugia il giorno 9 marzo 1926)', *Archeion* 7 (1926): 36–48.

—— *La storia della scienza in Italia. Saggio di bibliografia di storia della scienza* (Firenze: Libreria della voce, 1916).

—— *Lavoisier* (Genova: A.F. Formiggini, 1916).

—— *Le scuole ionica, pythagorica ed eleata (I prearistotelici, I.)* (Firenze: Libreria della Voce, 1916).

—— *Manuale di storia della scienza. Storia. Antologia. Bibliografia. – Antichità* (Roma: Leonardo da Vinci, 1925).

—— *Pagine di storia della chimica* (Roma: Leonardo da Vinci, 1922).

—— *Panorama general de historia de la ciencia* (Buenos Aires: Espasa-Calpe, 1945).

—— 'Patologia sessuale', *Archeion* 1/2 (1927): 81–94.

—— 'Pour l'organisation des historiens des sciences', *Archeion*, 9 (1928): 497–503.

—— 'Scienza e filosofia', *Rivista di filosofia* 2/5 (1910): 599–608.

—— 'Storia delle scienze; rassegna e note sulle pubblicazioni ad esse relative', *Rivista di filosofia* 4 (1912): 508–21.

—— 'Sul concetto di storia della scienza', *Rivista di storia critica delle scienze mediche e naturali* 7 (1916): 42–46.

—— 'Un viaggio in Germania', *Archivio di storia della scienza* 7 (1926): 343–81.

—— '[untitled]', *Archeion*, 8 (1927): 1–5.

—— (ed.), *Gli scienziati italiani dall'inizio del Medioevo ai nostri giorni* (Roma: Nardecchia, 1921).

Mieli, Aldo and Pierre Brunet, *Histoire des sciences. Antiquité* (Paris: Payot, 1935).

Milhaud, Gaston, *Leçons sur les origines de la science grecque* (Paris: Alcan, 1893).

—— *Les philosophes-géomètres de la Grèce. Platon et ses prédécesseurs* (Paris: Alcan, 1900).

Müller, Bertrand, 'Lucien Febvre et Henri Berr: de la synthèse à l'histoire-problème', in Agnès Biard, Dominique Bourel and Eric Brian (eds), *Henri Berr et la culture du XXe siècle: Histoire, science et philosophie* (Paris: Albin Michel/Centre international de synthèse, 1997), pp. 39–59.

Mus, Paul et al., 'La mythologie primitive et la pensée de l'inde', *Bulletin de la Société française de Philosophie* 37 (1937): 83–126.

Nabert, J., 'L'expérience interne chez Kant', *Revue de métaphysique et de morale* 31 (1924): 205–68.

N.D.L.R., 'Enseignement', *Revue de métaphysique et de morale* 3 (1895): 231–3.

Neri, Marina, 'Vers une histoire psychologique: Henri Berr et les Semaines internationales de synthèse (1929–1947)', in Agnès Biard, Dominique Bourel and Eric Brian (eds), *Henri Berr et la culture du XXe siècle. Histoire, science et philosophie* (Paris: Albin Michel/Centre international de synthèse, 1997), pp. 205–17.

Nisio, Francesco Saverio, 'Partecipazione come *scientia intuitiva*, Lévy-Bruhl e Spinoza', *Revue philosophique* 130 (2005): 323–33.

Pariente, Jean-Claude, 'Rationalisme et ontologie chez Gaston Bachelard', in Michel Bitbol and Jean Gayon (eds), *L'épistémologie française, 1830–1970* (Paris: Presses Universitaires de France, 2006), pp. 253–81.

Parinaud, André, *Gaston Bachelard* (Paris: Flammarion, 1996).

Parodi, Dominique, 'De l'explication dans les sciences par E. Meyerson', *Revue de métaphysique et de morale* 31 (1924): 585–97.

Parodi, Dominique, et al., 'La notion d'égalité sociale. Séance du 30 décembre 1909', *Bulletin de la Société française de Philosophie* 10 (1909): 53–79.

Pascal, Blaise, *Pensée et opuscules, publiés avec une introduction, des notices, des notes, par M. Léon Brunschvicg* (Paris: Hachette, 1897).

Paul, H.W., 'Scholarship and Ideology: The Chair of the General History of Science at the Collège de France, 1892–1913', *Isis* 67 (1976): 376–97.

Peano, Giuseppe, *Exemplo de Interlingua cum Vocabulario Interlingua Latino-Italiano-Français-Deutsch* (Cavoretto, Turin: Interlingua, 1913).

Petit, Annie, 'L'enseignement positiviste: Auxiliaire ou obstacle pour l'histoire des sciences?', *Revue d'histoire des sciences* 58 (2005): 329–65.

—— 'L'héritage du positivisme dans la création de la chaire d'histoire générale des sciences au Collège de France', *Revue d'histoire des sciences* 48/4 (1995): 521–56.

Pétrement, Simone, *La vie de Simone Weil*, vol. 1: *1909–1934* (Paris: Fayard, 1972).

Piaget, Jean et al., 'Les trois systèmes de la pensée de l'enfant. Séance du 17 mai 1928', *Bulletin de la Société française de Philosophie* 28/4 (1928): 97–141.

Piéron, Henri et al., 'La notion d'instinct. Séance du 28 mai 1914', *Bulletin de la Société française de Philosophie* 14 (1914): 301–36.

—— 'L'étude biologique de la mémoire. Séance du 12 janvier 1911', *Bulletin de la Société française de Philosophie* 11 (1911): 1–27.

Pierre-Duhem, Hélène, *Pierre Duhem, un savant français* (Paris: Plon, 1936).

Pinto, Louis, 'Le détail et la nuance; la sociologie vue par les philosophes dans la *Revue de métaphysique et de morale*, 1893–1899', *Revue de métaphysique et de morale* 98 (1993): 141–74.

Platania, Margherita, *Le parole di Clio: Polemiche storiografiche in Francia 1925–1945* (Napoli: Bibliopolis, 2001).

—— 'Le projet de *Vocabulaire historique*', in Agnès Biard, Dominique Bourel and Eric Brian (eds), *Henri Berr et la culture du XX*^e^ *siècle. Histoire, science et philosophie* (Paris: Albin Michel/Centre international de synthèse, 1997), pp. 231–9.

Pluet-Despatin, Jacqueline, 'Henri Berr éditeur. Elaboration et production de "L'évolution de l'humanité" ', in Agnès Biard, Dominique Bourel and Eric Brian (eds), *Henri Berr et la culture du XX*e *siècle. Histoire, science et philosophie* (Paris: Albin Michel/Centre international de synthèse, 1997, pp. 241–67.

Pogliano, Claudio, 'Aldo Mieli, storico della scienza (1879–1950)', *Belfagor* 38 (1983): 537–57.

Polizzi, Gaspare, *Tra Bachelard e Serres: aspetti dell'epistemologia francese del Novecento* (Messina, Armando Siciliano, 2003).

Prampolini, Camillo, *La predica di Natale* (Milano: Fondazione Giangiacomo Feltrinelli, n.d. [1897]).

Prochasson, Christophe, 'Philosopher au XXe siècle: Xavier Léon et l'invention du "système R2M" (1891–1902)', *Revue de métaphysique et de morale* 98 (1993): 109–40.

Prost, Antoine, *Histoire de l'enseignement en France* (Paris: Colin, 1968).

Pyenson, Lewis, 'Inventory as a Route to Understanding: Sarton, Neugebauer, and Sources', *History of Science* 33 (1995): 253–82.

Quality Assurance Agency for Higher Education website, 'Subject benchmark statements' http://www.qaa.ac.uk/academicinfrastructure/benchmark/honours/philosophy.asp (accessed 18 April 2006).

Rabinow, Paul, 'Introduction: A Vital Rationalist', in François Delaporte (ed.), *A Vital Rationalist: Selected Writings from Georges Canguilhem* (New York: Zone Books, 1994), pp. 11–22.

Rédaction, La, 'La vie du Centre. La section de synthèse générale. Son programme', *Revue de synthèse* 51 (1931): 127–31.

Redondi, Pietro, *Epistemologia e storia della scienza. Le svolte teoretiche da Duhem a Bachelard* (Milano: Feltrinelli, 1978).

—— 'Henri Berr, Hélène Metzger et Alexandre Koyré: la religion d'Henri Berr', in Agnès Biard, Dominique Bourel and Eric Brian (eds), *Henri Berr et la culture du XX*e *siècle. Histoire, science et philosophie* (Paris: Albin Michel/Centre international de synthèse, 1997), pp. 139–57.

Redondi, Pietro and P.V. Pillai (eds), *The History of Science: The French Debate* (London: Sangam Books, 1989).

Renard, Gilles, *L'épistémologie chez Georges Canguilhem* (Paris: Editions Nathan, 1996).

Renouvier, Charles, *Esquisse d'une classification systématique des doctrines philosophiques* (2 vols; Paris: Fischbacher, 1885).

—— *Uchronie (L'utopie dans l'histoire): Esquisse historique apocryphe du développement de la civilisation européenne tel qu'il n'a pas été, tel qu'il aurait pu être* (Paris: Bureau de la critique philosophique, 1876).

Rey, Abel, 'A propos de "l'explication dans les sciences" de M. Meyerson', *Revue de synthèse historique* 32 (1921): 123–40.

—— 'French Philosophy in 1926 and 1927', *The Philosophical Review* 37/6 (1928): 527–56.

—— 'Histoire de la science ou histoire des sciences', *Archeion* 12/1 (1930): 1–4.

—— *La philosophie moderne* (Paris: Flammarion, 1908).

—— 'La philosophie scientifique de M. Duhem', *Revue de métaphysique et de morale* 12 (1904): 699–744.

—— 'L'*a priori* et l'expérience dans les méthodes scientifiques', *Revue de métaphysique et de morale* 16 (1908): 883–8.

—— *La science dans l'antiquité, vol. 1: La science orientale avant le grecs* (Paris: La Renaissance du livre, 1930).

—— *La science dans l'antiquité, vol. 2: La jeunesse de la science grecque* (Paris: La Renaissance du livre, 1933).

—— *La science dans l'antiquité, vol. 3: La maturité de la pensée scientifique en Grèce* (Paris: La Renaissance du livre, 1939).

—— *La science dans l'antiquité, vol. 4: L'apogée de la science technique grecque: Les sciences de la nature et de l'homme, les mathématiques d'Hippocrate à Platon* (Paris: La Renaissance du livre, 1946).

—— *La science dans l'antiquité, vol. 5: L'apogée de la science technique grecque: L'essor de la mathématique* (Paris: La Renaissance du livre, 1948).

—— *La théorie de la physique chez les physiciens contemporains. Troisième édition revue et augmentée* (Paris: Alcan, 1930).

—— 'L'évolution de la pensée: De la pensée primitive à la pensée actuelle', in Lucien Febvre (ed.), *Encyclopédie française* (Paris: Société de Gestion de l'Encyclopédie Française, 1937), pp. 1.10.3–1.20.11.

—— 'Revue d'histoire des sciences et d'histoire de la philosophie dans ses rapports avec les sciences', *Revue de synthèse historique* 31 (1920): 121–35.

—— 'Sur le positivisme absolu', *Revue philosophique* 34 (1909): 65–6.

—— 'Vers le positivisme absolu', *Revue philosophique* 34 (1909): 461–79.

Rheinberger, Hans-Jörg, 'Gaston Bachelard and the Notion of "Phenomenotechnique"', *Perspectives on Science* 13/3 (2005): 313–28.

—— 'Reassessing the Historical Epistemology of Georges Canguilhem', in Gary Gutting (ed.), *Continental Philosophy of Science* (Oxford: Blackwell, 2005), pp. 187–97.

Richard-Foy, Emile, 'Einstein et sa conception d'un espace fini', *Revue philosophique* 49 (1924): 67–103.

Robin, Léon, *Aristote* (Paris: Presses Universitaires de France, 1944).

—— *La pensée grecque et les origines de l'esprit scientifique* (Paris: La Renaissance du livre, 1923).

—— *La théorie platonicienne de l'amour* (Paris: Alcan, 1933 [1908]).

—— *La théorie platonicienne des idées et des nombres. Etude historique et critique* (Paris: Alcan, 1908).

—— *Les rapports de l'être et de la connaissance d'après Platon. Cours professé en Sorbonne pendant l'année scolaire 1932–33, publié par Pierre-Maxime Schuhl* (Paris: Presses Universitaires de France, 1957).

—— 'L'histoire et la légende de la philosophie', *Revue philosophique* 60 (1935): 161–75.

—— 'Notice', in Platon, *Banquet* (Paris: Les Belles Lettres, 1992 [1929]).

—— 'Notice', in Platon, *Phédon* (Paris: Les Belles Lettres, 1983 [1926]).

—— 'Notice', in Platon, *Phèdre* (Paris: Les Belles Lettres, 1985 [1933]).

—— *Platon* (Paris: Alcan, 1935).

—— *Pyrrhon et le scepticisme grec* (Paris: Presses Universitaires de France, 1944).

Robin, Léon et al., 'Sur la notion d'histoire de la philosophie. Séance du 25 avril 1936', *Bulletin de la Société française de Philosophie* 36 (1936): 101–40.

Robinson, L., 'Contribution à l'histoire de l'évolution philosophique de Kant', *Revue de métaphysique et de morale* 31 (1924): 269–353.

Rose, Nikolas, 'Life, Reason and History: Reading Georges Canguilhem', *Economy and Society* 27/2–3 (1998): 154–70.

Roth, M., *Knowing and History: Appropriation of Hegel in Twentieth-Century France* (Ithaca and London: Cornell University Press, 1988).

Roudinesco, Elisabeth, *Histoire de la psychanalyse en France. 1: 1885–1939* (Paris: Fayard, 1994 [1982]).

—— *La Bataille de cent ans: Histoire de la psychanalyse en France. 2: 1925–1985* (Paris: Seuil, 1986).

Roussellier, Nicolas, 'Halévy (Elie)', in Jacques Julliard and Michael Winock (eds), *Dictionnaire des intellectuels français: les personnes, les lieux, les moments* (Paris: Seuil, 1996), pp. 582–3.

Ruyssen, T., 'Les origines kantiennes de la Société des Nations', *Revue de métaphysique et de morale* 31 (1924): 355–72.

Ruyssen, T. et al., 'Pacifisme et patriotisme. Séance du 28 novembre 1907', *Bulletin de la Société française de Philosophie* 8 (1907): 1–30.

Saint Anselm, *Fides quaerens intellectum. Introduction, texte, traduction par A. Koyré* (Paris: Vrin, 1927).

Saint-Sernin, Bertrand, 'Georges Canguilhem à la Sorbonne', *Revue de métaphysique et de morale* 90 (1985): 84–92.

Salvadori, Roberto, *Hegel in Francia: Filosofia e politica nella cultura francese del Novecento* (Bari: De Donato, 1974).

Sarton, George, *Introduction to the History of Science* (3 vols; Baltimore: William and Wilkins, 1927–47).

Schmaus, Warren, *Durkheim's Philosophy of Science and Sociology of Knowledge: Creating and Intellectual Niche* (Chicago and London: The University of Chicago Press, 1994).

—— 'Lévy-Bruhl, Durkheim, and the Positivist Roots of the Sociology of Knowledge', *Journal of the History of the Behavioral Sciences* 32/4 (1996): 424–40.

—— 'Renouvier and the Method of Hypothesis', *Studies in History and Philosophy of Science* 38 (2006): 132–48.

Schuhmann, Karl, 'Alexandre Koyré et les phénoménologues allemands', *History and Technology* 4 (1987): 149–65.

Schuhl, Pierre-Maxime, '[Hommage à Henri Berr]', *Revue de synthèse* 85 (1964): 135–36.

—— *Machinisme et philosophie* (Paris: Presses Universitaires de France, 1969 [1938]).

Schuwer, C., 'La pensée italienne contemporaine: L'idéalisme actuel: Giovanni Gentile', *Revue philosophique* 49 (1924): 82–123.

Schwartzkoppen, Maximilian von, *Les carnets de Schwartzkoppen: la vérité su Dreyfus, édité par Bernhard Schwartfeger et traduits sur le texte allemand par Alexandre Koyré; préface Lucien Lévy-Bruhl* (Paris: Rieder, 1930).

Section d'histoire des sciences, 'Communications officielles', *Archeion* 15 (1933): 153–60.

—— 'Communications officielles', *Archeion* 19 (1937): 402.

—— 'Séance du 19 mars 1930', *Archeion* 12 (1930): 375–79.

—— 'Séance du 23 janvier 1935', *Archeion* 17 (1935): 81–4.

—— 'Séance du 20 novembre 1935', *Archeion* 18 (1936): 75–9.

—— 'Séance du 22 janvier 1936', *Archeion* 18 (1936): 238–47.

—— 'Séance du 21 avril 1937', *Archeion* 19 (1937): 254–8.

Seignobos, Charles et al., 'L'inconnu et l'inconscient en histoire. Séance du 28 mai 1908', *Bulletin de la Société française de Philosophie* 8 (1908): 217–47.

Sergescu, Pierre, 'Aldo Mieli', *Actes du Congrès International d'Histoire des Sciences* 6 (1951): 79–95.

Sernin, André, 'Alain [Emile Chartier]', in Jacques Julliard and Michel Winock (eds), *Dictionnaire des intellectuels français: les personnes, les lieux, les moments* (Paris: Seuil, 1996), pp. 43–4.

—— *Alain: un sage dans la cité* (Paris: Robert Laffont, 1985).

Serres, Michel, 'La réforme et les sept péchés', *L'Arc* 42 (1970): 14–28.

Simon-Nahum, Perrine, 'Xavier Léon/Elie Halévy: Correspondance (1891–1898)', *Revue de métaphysique et de morale* 98 (1993): 3–58.

Sinaceur, Hourya Benis, 'From Kant to Hilbert: French Philosophy of Concepts in the Beginning of the Twentieth Century', in José Ferreirós and Jeremy Grey (eds), *The Architecture of Modern Mathematics* (Oxford: Oxford University Press, 2006), pp. 349–76.

Sirinelli, Jean-François, *Génération intellectuelle: khâgneux et normaliens dans l'entre-deux-guerres* (Paris: Presses Universitaire des France, 1994).

—— *Intellectuels et passions françaises: manifestes et pétitions au XXe siècle* (Paris: Gallimard, 1996 [1990]).

Société française de philosophie webpage http://www.sofrphilo.fr/ (accessed 1 June 2007).

Stock, Phyllis H., 'Students versus the University in Pre-World War Paris', *French Historical Studies* 7/1 (1971): 93–110.

Stoffel, Jean-François, *Bibliographie d'Alexandre Koyré* (Firenze: Olschki, 2000).

Susani, Carola, 'Una critica della norma nell'Italia del fascismo', in Enrico Venturelli (ed.), *Le parole e la storia: Ricerche su omosessualità e cultura* (Bologna: Il Cassero, 1991), pp. 110–19.

Tannery, Paul, 'De l'histoire générale des sciences', *Revue de synthèse historique* 8 (1904): 1–16.

Taton, René, 'Alexandre Koyré et l'essor de l'histoire des sciences en France (1933 à 1964)', *History and Technology* 4 (1987): 37–53.

Telkès-Klein, Eva, 'Emile Meyerson: A Great Forgotten Figure', *Iyyun* 52 (2003): 235–44.

—— 'Emile Meyerson, d'après sa correspondance. Une première ébauche', *Revue de synthèse* 125, 5th ser. (2004): 197–215.

Tijiattas, M., 'Bachelard and Scientific Realism', *Philosophical Forum* 22 (1991): 203–210.

Tiles, Mary, *Bachelard: Science and Objectivity* (Cambridge: Cambridge University Press, 1984).

Tolédano, André D., 'Henri Berr, Le maître et l'ami au jour le jour', *Revue de synthèse* 85 (1964): 144–7.

Tosi, Lucia, 'La trayectoria de Aldo Mieli en el Centre international de synthèse', *Saber y Tiempo* 4 (1997): 449–62.

Vadée, M., *Bachelard ou le nouvel idéalisme épistémologique* (Paris: Editions sociales, 1975).

Vandenbussche, R., 'Parti Radical', in Jean-François Sirinelli (ed.), *Dictionnaire historique de la vie politique française au XXe siècle* (Paris: Presses Universitaires de France, 2003), pp. 895–906.

Vendryès, J. et al., 'Le progrès du langage. Séance du 30 novembre 1922', *Bulletin de la Société française de Philosophie* 22 (1922) 151–69.

Vernant, D., 'Epistémologie', in Sylvain Auroux (ed.), *Encyclopédie philosophique universelle, vol. 2: Les notions philosophiques* (Paris: Presses Universitaires de France, 1990), p. 813.

Vinti, Carlo, *Il soggetto qualunque: Gaston Bachelard fenomenologo della soggettività epistemica* (Napoli: Edizioni scientifiche italiane, 1997).

Vogt, Paul W., 'Identifying Scholarly and Intellectual Communities: A Note on French Philosophy 1900–1939', *History and Theory* 21/2 (1982): 267–78.

Wahl, Jean, 'Commentaire d'un passage de la phénoménologie de l'esprit de Hegel', *Revue de métaphysique et de morale* 34 (1927): 441–71.

—— *Etude sur le Parménide de Platon* (Paris: Rieder, 1926).

—— *Le malheur de la conscience dans la philosophie de Hegel* (Paris: Rieder, 1929).

—— *Les philosophies pluralistes d'Angleterre et d'Amérique* (Paris: Alcan, 1920).

—— *Tableau de la philosophie française* (Paris: Gallimard, 1962).

—— *Vers le concret: Etudes d'histoire de la philosophie contemporaine* (Paris: Vrin, 1932).

Wallon, Henri, 'La mentalité primitive et celle de l'enfant', *Revue philosophique* 53 (1928): 82–105.

Weber, Louis et al., 'Liberté et langage. Séance du 23 juin 1921', *Bulletin de la Société française de Philosophie* 21 (1921): 75–106.

Wesseling, H.L., 'Commotion at the Sorbonne: The Debate on the French University, 1910–1914', *European Review* 9/1 (2001): 89–96.

Wilbois, J. et al., 'Une nouvelle position du problème moral. Séance du 2 janvier 1914', *Bulletin de la Société française de Philosophie* 14 (1914): 1–59.

Worms, Frédéric, 'Between Critique and Metaphysics: Science in Bergson and Brunschvicg', *Angelaki* 10/2 (2005): 39–57.

Zambelli, Paola, 'Alexandre Koyré alla scuola di Husserl a Gottinga', *Giornale Critico Della Filosofia Italiana* 78/3 (1999): 303–54.

—— 'Alexandre Koyré e Lucien Lévy-Bruhl. Dalle rappresentazioni collettive ai paradigmi del pensiero scientifico', *Intersezioni* 13 (1993): 395–409.

—— 'Alexandre Koyré, "Present Trends of French Philosophical Thought": Introduction', *Journal of the History of Ideas* 59/3 (1998): 521–30.

—— 'Alexandre Koyré versus Lucien Lévy-Bruhl: From Collective Representations to Paradigms of Scientific Thought', *Science in Context* 8/3 (1995): 531–55.

—— 'Filosofia e politica nell'esilio: Alexandre Koyré, Jacques Maritain e l'Ecole Libre a New York', *Giornale Critico Della Filosofia Italiana* 18/1 (1998): 73–112.

Index